Lecture Notes in Computer Sci

Edited by G. Goos and J. Hartmanis

Advisory Board: W. Brauer   D. Gries   J. Stoer

Frances Newbery Paulisch

The Design of an Extendible Graph Editor

Springer-Verlag

Berlin Heidelberg New York
London Paris Tokyo
Hong Kong Barcelona
Budapest

Series Editors

Gerhard Goos
Universität Karlsruhe
Postfach 69 80
Vincenz-Priessnitz-Straße 1
D-76131 Karlsruhe, Germany

Juris Hartmanis
Cornell University
Department of Computer Science
4130 Upson Hall
Ithaca, NY 14853, USA

Author

Frances Newbery Paulisch
Siemens AG, ZFE BT SE 32
Otto-Hahn-Ring 6, D-81739 München, Germany
E-mail: paulisch@ztivax.zfe.siemens.de

CR Subject Classification (1991): D.2.2, I.3.4, D.1.5, G.2.2, H.2.3, D.2.m

ISBN 3-540-57090-X Springer-Verlag Berlin Heidelberg New York
ISBN 0-387-57090-X Springer-Verlag New York Berlin Heidelberg

Typesetting: Camera ready by author
45/3140-543210 - Printed on acid-free paper

Preface

Graphs can be used to convey information about relationships in many applications. State transition diagrams, PERT/CPM charts, call graphs, and entity-relationship diagrams are a few examples of many applications involving graphs. Typically, nodes in the graph represent items in the application (e.g. a state, an activity, a program module) and the edges represent the relationships among these items (e.g. state transition, activity duration, procedure invocation). A graph editor is an interactive tool that presents a graph to the user pictorially and allows the user to edit the graph. The recent proliferation of graph editors for particular applications indicate their effectiveness as the graphical user interface to an application. Many designers, however, are hesitant to use the graph editor model because of the high cost of developing such a graphical user interface.

This book presents the design of an extendible graph editor, which is a graph editor that can be adapted easily to many different application areas. The advantages of using a graph editor will thus be available for a minimal customization effort. Several fundamental and recurring problem areas associated with graph editors are investigated and a solution is proposed for each. The specific topics investigated are:

- Graph layout: How can application-specific layout requirements, individual preferences, and layout stability be integrated with layout algorithms? A layout constraint mechanism is presented which can easily be combined with various graph layout algorithms.

- Graphical abstraction: How can users deal with large graphs containing hundreds of nodes and thousands of edge crossings? A novel clustering technique called *edge concentration* is presented which can reduce the apparent complexity of the graph. Alternatively, a subgraph can be specified and viewed as a multi-level graphical abstraction either in the context of the graph or in a separate view.

- Persistence: How can the graph structures produced by the editor be kept in long-term storage, especially if the node and edge data structures have been extended for a particular application? The proposed solution uses a standardized, external format for graphs. A program generator tool reads the graph, node, and edge class declarations and automatically generates routines for reading, writing, and editing these data structures.

- Extendibility: How should the editor kernel be structured to be adaptable to various applications? The object-oriented design of the proposed graph editor makes it easy to adapt.

To demonstrate their feasibility, the proposed solutions have been incorporated into EDGE, an extendible graph editor prototype. EDGE has been

VI

adapted to a number of applications including: a browser for entity-relationship diagrams, a tool for visualizing software configurations, a PERT chart editor, a call graph animator, a directory editor, and a logic simulator.

This document is a revised version of my doctoral dissertation from the Faculty of Informatics at the University of Karlsruhe (FRG) presented on May 7, 1991. I thank my advisor, Prof. Dr. Walter F. Tichy, for his invaluable guidance over the past years as well as my dissertation's co-referee, Prof. Dr. A. Schmitt.

I particularly thank Karl-Friedrich Böhringer, Stefan Manke, and Stefan Strugies for their valuable contributions to EDGE. Special thanks go to Bala Krishnamurthy, without whose encouragement I might never have made it this far. The support of my family was an important factor in completing this work.

Finally, I would like to thank the EDGE users at companies, research institutions, and universities in the United States and Europe. Their strong interest lends credence to my claim that an extendible graph editor is an appropriate graphical user interface to a wide range of applications.

Frances Newbery Paulisch
June 1993

Contents

List of Figures

List of Tables

Chapter 1

Introduction

The recent proliferation of high quality graphics workstations has been closely followed by interactive tools that present information to the user graphically rather than using traditional, textual representations. A graphical user interface makes tools easier to learn, use, and understand because humans recognize patterns better when they are presented pictorially. In [Rob87], Robins gives a compelling example of why "a picture is worth a thousand words". Here, two representations of a graph are given – one as a list of edges and the other as a drawing of the graph (see figure 1.1[1]). Important properties of the graph – that it is a binary tree and that "K" is the root of the tree – are immediately obvious from the drawing. The list of edges contains the same information, but the user has to consider each edge and compute the transitive closure (possibly sketching a drawing in the process) to extract this information.

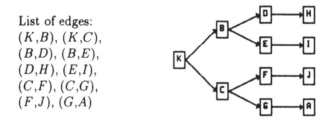

List of edges:
(K,B), (K,C),
(B,D), (B,E),
(D,H), (E,I),
(C,F), (C,G),
(F,J), (G,A)

Figure 1.1: Textual and pictorial representations of a graph

There are many different ways of presenting information graphically, and one of the most general is to represent the information as a graph. Informally, a

---

[1]Unless otherwise noted, the graphs shown as examples are drawn by the EDGE graph editor described in chapter 8.

graph consists of a set of nodes and a set of edges. Each node typically represents some object and the edges represent binary relationships between these objects. Information can be associated with the nodes and edges of the graph. Graphs are used to convey physical or conceptual information in many different application areas. The following lists a few applications of graphs in computer-related fields:

- **Software**: Graphs are used in all phases of software development from flowcharts, data structure animation, data flow diagrams, finite state automata, petri nets, and syntax graphs to call graphs and software configuration dependency graphs. They are also heavily used in the relatively new area of visual programming [Shu89, Gli90].

- **Hardware**: Computer hardware gates can be interconnected to form combinational logic networks.

- **Database**: An entity-relationship diagram [Che76, Gan90], commonly used for the conceptual design of database schemas, is a graph consisting of entities, relations, and attributes. The user interface of hypertext systems [Con87] is often based on graphs.

- **Networking**: Graphs are used to display network configurations where nodes represent machines and edges the physical connection between them. Reachability graphs are used to verify communication protocols[CL88].

- **Artificial Intelligence**: Semantic nets used to represent knowledge[Bra79].

- **Business**: PERT and CPM charts [CCP87], used in the area of project management, are graphs that help a project manager visualize the dependency relationships among various subprojects.

The terminology used in the application areas listed above indicates a wider variety than is actually the case. Syntax trees, entity-relationship diagrams, networks, semantic nets, PERT charts and the rest – all are different forms of graphs.

Just as graphs provide a general representation of information, editing is a general model of interaction for user interfaces. In [DS90] it is argued that any interactive application could present a graphical representation of its data and allow the user to edit it and to update the representation. For example, a mail program could present a graphical representation of a mailbox which the user would edit to read or send mail messages. *Direct manipulation* [Shn83], is a particular form of interaction in which the user specifies objects by selecting them "directly" on the screen using a pointing device (e.g. a mouse) rather than specifying them "indirectly" (e.g. by name).

A *graph editor* is an interactive tool that presents a graph to the user pictorially and allows the user to edit the graph. The user can add, delete, or edit

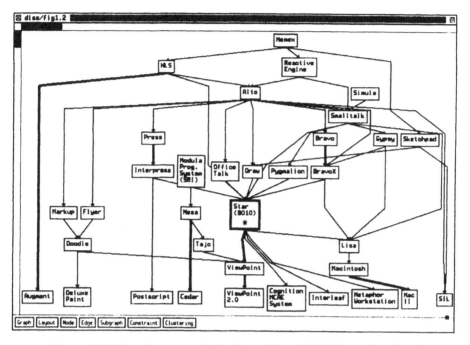

Figure 1.2: Development of the Xerox Star (EDGE graph editor)

nodes and edges in the graph and the changes will be reflected in the display of the graph. A graph editor is a powerful and widely-applicable tool because it combines a general graphical representation of information (a graph) with a general model of interaction (an editor).

Graph editors can support a wide range of user interaction. Figures 1.2 and 1.4 show examples of two extremes – one with little or no editing of the graph and the other with frequent and continuous updates.

The graph editor shown in figure 1.2 depicts the development of the Xerox Star, a personal computer designed for use by business professionals in an office environment. This information was extracted from a (presumably manually-drawn figure) given in [JRV⁺89] and shown in figure 1.3. The graph shows how related systems influenced each other (the thick lines represent direct successors of a system). In this example, the information being displayed is relatively static and little or no editing of the graph is required. The placement of the nodes and edges in figure 1.2 is done automatically as opposed to being positioned manually by the user. The layouts are of comparable quality. Juding by one of the often-used concrete measures of layout quality, the number of edge crossings, the layouts are equally good (both have 24 crossings). However, reliance on the automatic layout of the graph (which takes 15 seconds on a Sun 3 workstation) is surely faster than a manual layout. This example shows the benefit of using

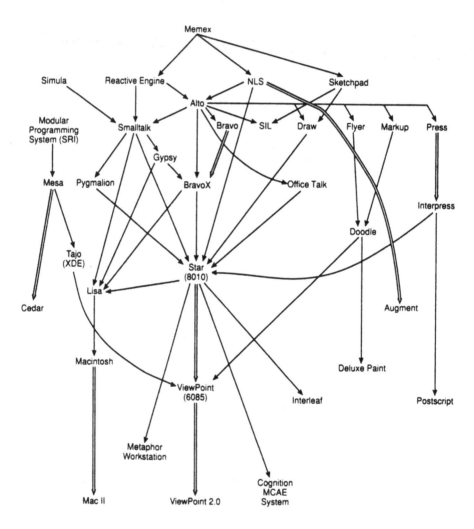

Figure 1.3: Development of the Xerox Star (from [JRV89])

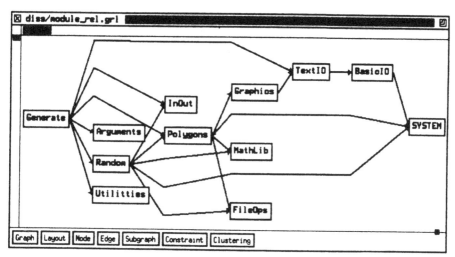

Figure 1.4: Graph editor depicting the import/export relations between modules

a graph-based tool to present information to the user, and, in particular, the benefits of automatic graph layout.

At the other extreme, the information being displayed may be changing rapidly. Consider the graph editor shown in figure 1.4 depicting the import/export relationship among modules of a program [Luc90]. The graph in this case is automatically generated from the source code of a Modula program and thus the frequency of changes is on the order of minutes rather than years. Conceivably, the user could access the source code through the editor and changes would be reflected in the graph's representation immediately.

This dissertation presents the design of an *extendible graph editor* which is a graph editor that can easily be adapted to a wide variety of applications. Changes made by the user will not only be reflected in the graphical representation of the graph, but also in the application itself. The next section presents the motivation and goals of this work and points out the shortcomings of existing graph editors. The subsequent section presents an overview of the main research contributions of the thesis. This chapter closes with an overview of the organization of the rest of this presentation.

1.1 Motivation and Goals

Graph editors have been developed for numerous applications [BNT86, WP86, CL88, BCL90, Bru88, RDLK90]. When using a graph editor for a particular application, application-specific actions are associated with the editing of the graph. For example, consider a graph editor for project management which displays a PERT chart and recalculates and highlights the critical path after each editing command. The project manager using such a tool receives immediate visual feedback after changing the duration of one of the subprojects and thus is likely to gain a better understanding of the dependencies among the project parts than if the immediate visual feedback were not available.

The increasing use of graph editors for building graphical user interfaces to applications indicates their effectiveness. There are, however, several fundamental problems with graph editors that must be resolved before their use becomes more widespread. These include:

- **Graph layout**: It is important to present the user with a "nice" drawing of the graph from among the many possible graphical representations of a graph. An important aspect of a graph's appearance is the positioning of the nodes and edges of the graph. This is called the *graph layout*. One solution is to let the user position the nodes and edges of the graph manually. However, for large graphs this approach is too error-prone and tedious to be useful. Instead, a *layout algorithm*[2] can be used to place the nodes and edges according to one or more *layout aesthetics*. The layout aesthetics describe what attributes are important for producing a "nice" layout. For example, one of the common layout aesthetics is to minimize the number of edge crossings in the graph's layout. Often the layout aesthetics are chosen so that structural properties such as planarity, symmetry, or hierarchy are easily recognized. It is sometimes possible to satisfy several different layout aesthetics simultaneously as shown in figure 1.1 where the layout aesthetics for all three of these structural properties were satisfied.

  Automatic graph layout is the only reasonable approach for large graphs, but it has two major drawbacks. Firstly, it is difficult to take user- or application-specified constraints on the layout into account. A layout algorithm is concerned primarily with satisfying layout aesthetics related solely to the structure of the graph. The layout algorithm typically has no information concerning the semantics of the graph. Such information, for example that a particular set of nodes should always be placed together, must be provided by the user or the application. A second problem with automatic layout is that it is *instable*, meaning that a small change in the graph's structure can drastically change its layout. This can be frustrating to users because they temporarily lose their orientation in the graph.

- **Graphical abstraction**: The graphs used in realistic applications (with hundreds or thousands of nodes) are often so large that they become unmanageable. However, a graphical representation of the graph is only useful if the portions the user wants to see are on the screen and if the

---

[2]Unless otherwise noted, the terms *layout algorithm* and *graph layout* refer to the automatic placement of nodes and edges as opposed to their manual placement.

graph's representation is understandable. A user may need to see several views, possibly depicting different relations simultaneously, to take full advantage of the interactive facilities. Abstraction mechanisms are necessary to reduce the apparent complexity of the graph. By using these, the user can abstract away irrelevant details and concentrate on the portions of the graph that are of interest. This dissertation is not solely concerned with the *representation* of abstractions, but also with their *definition*. Ideally, the system would detect good candidates for graphical abstraction automatically. Most graph editors, however, do not offer sufficient abstraction mechanisms and are especially lacking in their automatic definition.

- **Persistence**: *Persistence* is the preservation of data structures beyond the execution of the program that created them. A graph editor must be able to save graphs in long-term storage (e.g. in a file on disk). Most graph editors only offer a simple external format consisting of a list of nodes and edges to save the graph. A graph editor offering persistence is able to save not only the basic structure of the graph, but many other attributes associated with it as well. This includes information about how the nodes, edges, and graph should be displayed as well as the status of the current editing session.

- **Extendibility**: A significant portion of the code of an interactive application consists of the graphical user interface[BMS86, SS78]. Direct manipulation interfaces, such as one uses in a graph editor, are among the most difficult kinds to implement [Wil83]. Unfortunately, the design of many currently available graph editors restricts their use to a particular application. Because the graphical user interface is expensive to develop, it is beneficial to develop a graph editor which can be adapted to various applications rather than building a new editor for each application. An *extendible graph editor* is a graph editor which can be adapted easily to many different applications. An extendible graph editor allows the application developer to customize the visual appearance of the graph (e.g. taking traditional representation of graphs in that application area into account) and provides mechanisms for interacting with the application (e.g. performing application-specific actions when a node or edge is selected).

The main reason for making a graph editor extendible is to increase the reusability of the user interface software. This also has the additional advantage of providing a consistent user interface across several applications.

1.2 Research Contributions

This book is concerned with the design of an extendible graph editor kernel that solves the aforementioned fundamental and recurring problems for graph editors.

Research from each of these problem areas is presented and solutions proposed for each. In particular, the main research contributions are:

- **Layout constraints and stability** : The proposed solution is to provide a layout constraint manager and an automatic layout algorithm which interact with one another. This allows user- and application-specific layout constraints to be taken into account when the graph is laid out. Layout stability is achieved by automatically generating layout constraints that record the current layout of the graph and then relaxing these constraints near the subgraph affected by the change. This restricts dramatic changes in the layout to a portion of the graph.

- **Edge concentration** : An automatic edge clustering mechanism is proposed that identifies complete subgraphs and replaces them with an alternative representation. The resulting graph has fewer edges and often fewer edge crossings. Two further graphical abstraction mechanisms, the hierarchical subgraph abstraction and the separate view, are also presented.

- **Extendible external format for graph editors** : A language for representing all information associated with a graph editor, including the representation of application-specific attributes, is proposed. This language can be used both as the input representation for a graph as well as to achieve persistence.

- **Program generator tool**: A program generator tool is presented which partially automates the customization of the graph editor to a particular application. This tool reads in the data structure declarations provided by the application developer and automatically extends the source code for the input, output, and menu routines by the application-specific attributes.

In all cases, the solutions rely on techniques which support (at least partial) automation, yet are still under control of the user.

1.3 Organization

Chapter 2 presents definitions from the areas of graph theory and graphical user interfaces which will be used throughout this dissertation. Although some definitions (e.g. for graph editor) may appear elsewhere as well, they will be given here for the sake of completeness. Chapter 3 presents related work in the area of graph editors. A wide range of existing graph editors is presented with particular emphasis on their contributions in the areas of graph layout, graphical abstraction, persistence, and extendibility.

Chapters 4 through 7 contain the main research contributions. Each of the aforementioned problem areas is treated in a separate chapter. Because these areas are so diverse, a section on related work will be included in each chapter rather than be merged into a single chapter on related work. The structure of each of these chapters is: presentation of related work, a description of the problem, the proposed solution, example(s), and a summary of the results. Chapter 4 covers the topic of graph layout including layout constraints and layout stability. Chapter 5 is on graphical abstraction. Chapters 6 and 7 are intertwined because they both concern the graph representation language GRL which is used as an external representation of a graph editor's data. Chapter 6 presents the standard version of the language and shows how it can be used to achieve persistence. Chapter 7 shows how this language can automatically be extended and presents the object-oriented design of an extendible graph editor.

To demonstrate their feasibility, the proposed solutions have been incorporated into EDGE, an extendible graph editor prototype. EDGE has been adapted to a number of applications including: a browser for entity-relationship diagrams, a tool for visualizing software configurations, a PERT chart editor, a call graph animator, a directory editor, and a logic simulator. Chapter 8 presents the EDGE graph editor. A description of each of the existing applications is given, including detailed descriptions of the application developer's effort to adapt EDGE to the applications. Finally, chapter 9 summarizes the results and suggests directions for future research. Chapter 10 presents conclusions that are drawn from the design and implementation of the extendible graph editor described here. Appendix A describes the lexical conventions used and gives an extended BNF for the grammar of GRL.

Chapter 2

Definition of Terms

2.1 Graphs

A *graph* $G = (N, E)$ is a set $N = N(G)$ of nodes and a set $E = E(G)$ of edges where each edge is an unordered pair of distinct nodes in N^1. By definition, a graph does not contain a *loop*, an edge from a node to itself; nor does it contain *multiple edges*, edges between the same pair of nodes.

If the ordering of the nodes (u, v) in edge $(u, v) \in E$ is irrelevant, then the edge is an *undirected edge*. If all edges of a graph are undirected then the graph is called an *undirected graph*. If there is an ordering of the nodes (u, v) in edge $(u, v) \in E$, then the edge is a *directed edge*. The node u is called the *source node* and v is called the *target node*. Furthermore, u is called the *predecessor* of node v and v is called the *successor* of u. If all edges of a graph are directed then the graph is called a *directed graph*.

The *degree* of a node is the number of edges incident at that node. For directed graphs there is a distinction between the *in-degree*, the number of incoming edges, and the *out-degree*, the number of outgoing edges of a node.

A *path* from n_1 to n_k is an ordered list of distinct nodes $n_1, n_2, ... n_k$ such that $(n_1, n_2), (n_2, n_3), ... (n_{k-1}, n_k) \in E$. A *cycle* is a path such that the beginning and ending nodes are the same. A graph containing a cycle is called a *cyclic graph*. An *acyclic graph* is a graph that does not contain any cycles.

A graph is *connected* if, for every pair of nodes x, y, there exists a path from x to y. Otherwise the graph is said to be *disconnected*. A directed graph is said

---

[1] Another commonly used name for "node" is "vertex" and another commonly used name for "edge" is "arc".

to be connected if its corresponding undirected graph is connected. A *tree* is a connected graph such that for any edge $(u, v) \in E$, the graph $G - (u, v)$ is disconnected. The *root node* of a directed tree is a node with in-degree zero.

Graph $G' = (N', E')$ is a *subgraph* of $G = (N, E)$ if $N' \subset N$ and $E' \subset E$.

A *planar graph* is a graph that can be embedded in the plane such that no edges cross. Each connected region of the planar graph is called a *face*.

A *bipartite graph* $G = (N1, N2, E)$ is a graph whose nodes can be partitioned into two subsets $N1$ and $N2$ ($N(G) = N1 \cup N2$ and $N1 \cap N2 = \emptyset$) and each edge joins a node in $N1$ with a node in $N2$. Similarly, a *tripartite graph* $G = (N1, N2, N3, E)$ is a graph whose nodes can be partitioned into three subsets $N1$, $N2$, and $N3$ ($N(G) = N1 \cup N2 \cup N3$ and $N1 \cap N2 \cap N3 = \emptyset$) and no edge joins nodes within the same node subset. A *complete graph* is a graph such that each node is connected with every other node. A graph $G' = (N1, N2, E')$ is a *complete bipartite subgraph* of $G = (N, E)$ if $N1 \subset N$, $N2 \subset N$, $E' \subset E$ and $E = N1 \times N2$.

A graph $G1 = (N1, E1)$ is said to *cover the edges* of graph $G2 = (N2, E2)$ if every edge $(u, v) \in E2$ is also $(u, v) \in E1$. A *partition* of the edges of a graph $G = (N, E)$ is a set of sets of edges E_1 $E_2, ... E_k$ such that each edge $e \in E$ is in exactly one E_i ($1 \leq i \leq k$).

A *topological ordering* of an acyclic directed graph is an ordering of the nodes in the graph such that node x will precede node y in the ordering if there is a path from x to y in the graph. This ordering is used to arrange the graph such that all (directed) edges point in the same direction. If special attention is paid to cycles, a topological ordering can also be applied to cyclic graphs. In this case, an edge which points in the non-standard direction is called a *backedge*.

During the topological sort, the nodes are partitioned into a set of *levels* such that all nodes on level i precede a node in level j ($i < j$). Various algorithms can be used to assign level numbers to a node. A commonly used algorithm is based on the recursive definition:

$$level(x) = \begin{cases} 0 & \text{if } x \text{ has no predecessors} \\ \max_{y \in \text{pred. of} x}\{level(y)\} + 1 & \text{otherwise} \end{cases}$$

An alternative assignment of level numbers places nodes with no predecessors at a higher level in the graph. For example, in the graph shown in figure 2.1, the node F may be assigned level 0 (as shown in figure 2.1(a)), but could also be assigned levels 1 or 2 (as shown in figure 2.1(b)). The *rank* of a node is defined to be the level number plus one (i.e. a root node has level 0, rank 1). The *level hierarchy* of a graph[STT81] is a partitioning of the nodes into subsets $N_1, N_2, ..., N_n$ ($N(G) = N_1 \cup N_2 \cup ... \cup N_n$ and $N_i \cap N_j = \emptyset$) where N_i consists of all nodes on level i of the graph. Every edge $(u, v) \in E$, where $u \in N_i$ and $v \in N_j$, satisfies $i < j$. A graph is called *hierarchized* if level numbers have

been assigned to all of the nodes. A hierarchy is called *proper* if the edges can
be partioned into subsets $E_1, E_2, ...E_{n-1}$ ($E(G) = E_1 \cup E_2 \cup ... \cup E_{n-1}$ and
$E_i \cap E_j = \emptyset$) such that $E_i \subset N_i \times N_{i+1}, i = 1, ...(n-1)$ (i.e. that the source and
target nodes of an edge are in adjacent levels).

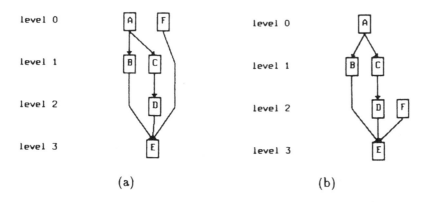

Figure 2.1: Level hierarchy of a directed acyclic graph

2.2 Graph Editors

A *graph editor* is an interactive tool that presents a graph to the user pictorially
and allows the user to edit the graph. A *graph browser* is like a graph editor,
but offers no editing capabilities. An *extendible graph editor* is a graph editor
which can be adapted easily to many different applications.

A graph editor usually runs on a machine that provides a bitmap screen and
a pointing device. A *bitmap screen* is a screen made up of a dense raster (ca.
1000 x 1000) of *pixels* which can each take on a color value. A *pointing device*
is a hardware device that continuously returns a pair of x- and y-coordinates
indicating a position on the screen. A *cursor* is a small symbol which tracks the
pointing device's position on the screen. Pointing devices typically also have
one or more *buttons* which the user can *click* to select the object currently being
pointed to. One of the most common pointing devices is a *mouse*. Rather than
pointing to the object directly on the screen, the user moves the mouse on a
flat surface near the machine and the cursor moves accordingly. Graphical user
interfaces in which the user selects objects on the screen using a pointing device
are called *direct manipulation* interfaces [Shn83].

The user interface of such graph-based tools provide facilities for viewing
large graphs. Typically there is a large scrollable region in which the graph
is displayed. In addition to a scrolling mechanism, other techniques (such as
a *layout overview* showing a small drawing of the entire graph and indicating

the current position using a *placemarker* box) may be available to help the user "navigate" through the graph. A *graph object* is an entity which the user can select, i.e. a node, edge, or subgraph. Each graph object has a set of *attributes* associated with it. These are typically used to control the display of the object (e.g. to specify its color or font), but in an extendible graph editor they may also serve an application-specific purpose. The user can examine the attributes of graph objects interactively. Through the user interface, the user can select graph objects and perform operations on them. In the case of a graph editor, the user can add, delete, or edit graph objects. In the case of an extendible graph editor, an explicit *application interface* defines the set of operations on the graph objects. Changes in the graph will be reflected in the display of the graph.

The use of an extendible graph editor as the user interface of an application usually involves several people. The designer of the graph editor kernel is called the *graph editor designer*. The *application developer* is the person who customizes the graph editor for a particular application. The application itself is written by the *application programmer*. The *user* is the person who actually uses the finished product. Naturally, each of these roles may actually be performed by a group of people and some persons may perform several roles.

Chapter 3

Related Work: Graph Editors

This chapter describes several interactive tools for displaying graphs. They are grouped into the categories special-purpose graph editors, general-purpose graph editors, and extendible graph editors.

- A *special-purpose graph editor* is a graph editor that is designed to work for one particular application only.

- A *general-purpose graph editor* allows the user to edit graph structures, but there is no application interface provided.

- An *extendible graph editor* provides both a general graph editor and an application interface. Editing operations performed on the graph will invoke the appropriate application procedures.

A short description is given for each system followed by a list of each system's relevance to the areas of graph layout, graphical abstraction, persistence, and extendibility.

Although one can produce drawings of graphs using an interactive drawing program, two essential features are lacking between a drawing program and a graph editor, and this limits their relevance in this presentation. Firstly, an interactive drawing program does not have a graph as the underlying data structure and therefore it treats the objects as independent symbols. Therefore, when a user moves a node, the attached edges do not automatically follow. Furthermore, because there is no underlying structure, there is no information to base an automatic layout on. When using an interactive drawing program, the user must place the nodes and edges manually, with the exception that the system may be able to make minor adjustments to "tidy up" the layout. Secondly, there are no facilities for interfacing to an application. On the positive side, interactive drawing programs do offer some facilities for graphical abstraction (e.g. allowing the user to set "links" to other "pages" in the document or to scale the document to an arbitrary size) and persistence (allowing the user to restart the system from a saved state). Despite the fact that they do not support interactive editing, two graph drawing systems (DRAG and DAG) are included in this chapter because of their significant contributions in the areas of graph layout and persistence.

3.1   Special-Purpose Graph Editors

GINCOD [BNTT85] is a graph editor for the conceptual design of database applications.

- **Graph layout:** It displays an entity-relationship model using an automatic layout algorithm [TBT83] designed especially for entity-relationship diagrams. The drawing area for the graph is split into rows and columns and each cell can display at most one node. The layout uses a grid standard and a prioritized list of layout aesthetics: minimize crossings, minimize bends, minimize total length of the edges, and minimize area of the diagram. The layout also takes one layout constraint into account: it will place a specified node in the center of the drawing. For each editing command, the user can specify whether automatic or manual layout is preferred.

- **Graphical abstraction:** The user may "expand" one of the nodes by replacing it either with an existing subgraph chosen from a database or by interactively creating a subgraph. The user must specify manually how the new subgraph should be connected with the remaining graph.

- **Persistence:** GINCOD is the graphical user interface to the INCOD system, a tool for defining data, transactions, and events using an entity-relationship model. INCOD also has a textual interface and the user can switch easily from one interface to the other. Persistence is achieved through the INCOD system. The graph representation displayed by GINCOD is produced by a "schema-to-diagram" translator tool.

- **Extendibility:** None.

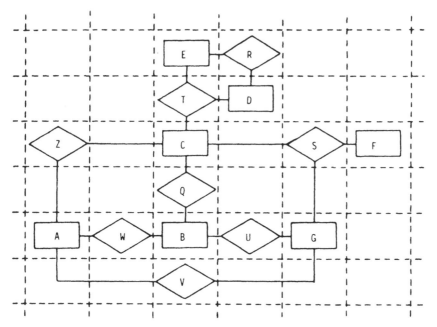

Figure 3.1: GINCOD editor (from [TBT83])

Software Through Pictures [WP86](a.k.a. IDEtool) is a set of graph editors designed to support the computer aided design of software. It provides entity-relationship, dataflow diagram, data structure, structure chart, and transition diagram editors. All of these graph editors have a consistent user interface and all interact with a common project database which is a repository for all project-related information. This set of tools can be used to check for consistency, generate code skeletons and prototypes, and generate hardcopy pictures in PIC [Ker82] or PostScript [Sys85] format.

- **Graph layout:** No automatic layout or abstraction mechanisms are provided.

- **Graphical abstraction:** None.

- **Persistence:** Persistence is achieved by saving information in the project database. The advantage of this approach is that the same database is available for all of the graph editors. Therefore, information that is saved by one tool may be used later by a different tool. The database also provides support for version control and configuration management.

- **Extendibility:** The user may extend the database schema and add new attributes to the tools, but the use of the tool is still restricted to the computer aided design of software.

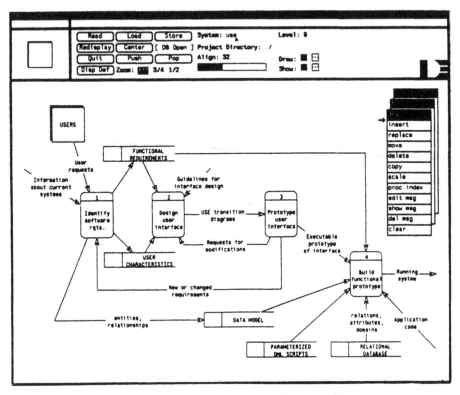

Figure 3.2: Software Through Pictures editor

PROSPEC [CL88] is a set of tools for interactively designing and verifying communication protocols. A single graph editor is used to display graphs representing the protocol topology, the machines, and the reachability graphs used to verify the protocol. A graphical user interface displays the reachability graph of the protocol and allows the user to examine its behavior. Problem areas such as deadlocked states are brought to the user's attention and the user can highlight the currently executing nodes or edges – thus making it easier for the user to debug the protocol than if only a textual representation were available.

- **Graph layout:** Although the graphs can be very large, no automatic layout facilities are available.

- **Graphical abstraction:** Application-specific abstraction mechanisms are used to reduce the complexity of the large (possibly infinite) reachability graphs. The user can choose to view the graph incrementally. In that case, the user controls which portions of the graph should be explored further.

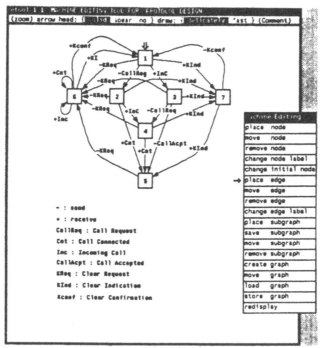

Figure 3.3: PROSPEC editor

- **Persistence:** For communication between the tools, a standard internal representation of the graph is used.
- **Extendibility:** None.

ParaGraph [BCL90] is a graph editor designed to support massively parallel programming environments. It is based on an extended form of aggregate rewriting graph grammars. ParaGraph allows the specification of families of graphs as opposed to individual graphs. The user of ParaGraph starts with the smallest member of the graph family and then applies a set of transformations to convert this graph to the next larger family member. This corresponds directly to the graph grammar and its productions.

- **Graph layout:** For small graphs, ParaGraph uses local automatic layout which replaces a node by a subgraph and scales the graph to fit around the expanded node. For more complex graphs, ParaGraph uses a heuristic based on the spring embedder algorithm [Ead84, KK89] (see section 4.1.2).
- **Graphical abstraction:** None.
- **Persistence:** None.

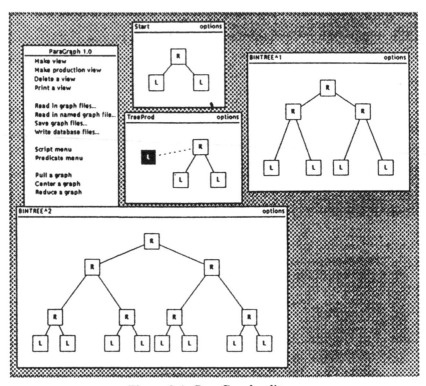

Figure 3.4: ParaGraph editor

- **Extendibility:** None.

GERM [Bru88] is a graph editor for browsing and editing databases. It has been used in the following application areas – to experiment with design representations, as a generic user interface for prototype development, and as a standalone system for database browsing.

- **Graph layout:** The layout of the graph is currently done manually, although the developers plan to use automatic graph layout algorithms in the near future.

- **Graphical abstraction:** Two abstraction mechanisms called "collection" and "aggregate" are available. A *collection* is a subgraph that can be operated on as a unit. It can either be an arbitrary set of nodes and edges selected by the user or it can be a set of nodes and edges that are logically related according to a definition given by the user. To specify such a definition, the user gives an entity name and a set of relation names. All entities and relations reachable from the root entity via the specified relations will be included in the collection. For example, a tree node and all of its children could be defined as a collection. An *aggregate* is a mechanism to display a subgraph as a single node. The subgraph may either be a collection or an arbitrary

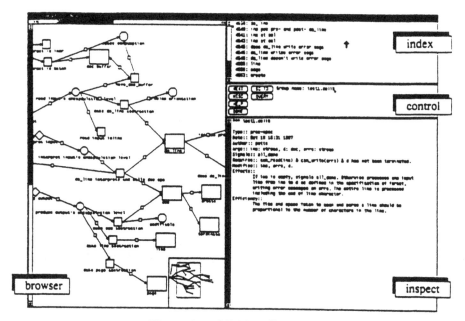

Figure 3.5: GERM editor

subgraph selected by the user. Aggregates are recursive (i.e. they may contain other aggregates). Aggregates are non-disjoint, meaning that nodes and edges may be shared between aggregates. The user can choose whether aggregates should be displayed as a single node or as the subgraph they represent.

A layout overview of the entire graph is displayed to provide the user with a global view of the graph.

- **Persistence:** A GERM user specifies a schema in a file. This schema includes an entity-relationship schema, a list of display information for the nodes and edges, and definitions for the abstraction mechanism. This schema is translated into a set of type definitions and the graph is stored in a database.

- **Extendibility:** GERM can be customized by changing definitions in the schema. However, GERM has a fixed set of operations and it is not easy to change them to an application-specific operation.

VIFOR [RDLK90] is a graph editor designed to assist in the maintenance of Fortran programs. In addition to viewing the source code, the user can see a graph representing the code. The user selects the information that is displayed in the graph. Possible choices for nodes are modules, include files, subprograms, and global data declarations. Possible choices for relations are belongs-to (e.g. a declaration belongs to a module), call, and reference.

Figure 3.6 shows a graph of a small program. The left column lists the
subprograms, and the right column shows the global data declarations.
The arrows to the left of the lefthand side column show the calls relation
and the other arrows show the reference relation.

- **Graph layout:** The graph layout in this two-column format is done
  automatically. This simple layout algorithm is sufficient for this particular application, but is not applicable to graphs in general. One of
  the advantages of this layout is that the addition of new nodes will
  not dramatically affect the layout (i.e. the layout is relatively stable).

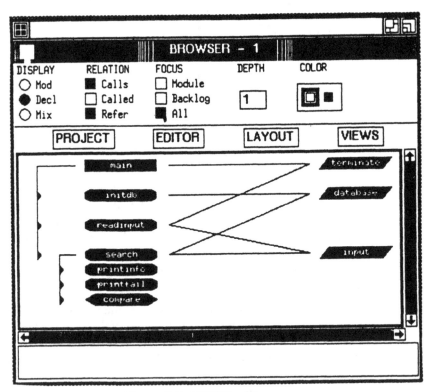

Figure 3.6: VIFOR editor

- **Graphical abstraction:** VIFOR offers a simple abstraction mechanism called "views". The user can restrict the set of nodes and edges
  by making appropriate selections in the menu. The view can either
  be shown in a separate window or can be appended to the existing
  view. Although this facility does not enable the user to see how the
  subgraph currently being viewed fits into the overall system, it does
  give the user the ability to concentrate on a particular subgraph of

interest.

- **Persistence:** There is no explicit support for persistence, but one can view the source code itself as a persistent external representation.

- **Extendibility:** None.

3.2 General-Purpose Graph Editors

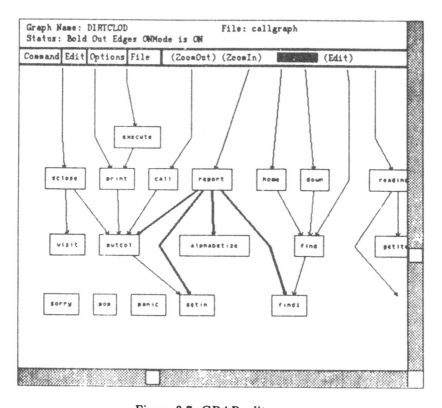

Figure 3.7: GRAB editor

GRAB [RDM+87, Dav86](**GRA**ph Browser) refers to itself as a graph browser, but in the defined terminology it is actually a graph editor, because it does support graph editing. One of the GRAB applications is a program browser(see figure 3.7). The list of nodes and edges used as input is generated from a call graph generator that examines the source code. Selecting

a node causes an editing session on the file containing that procedure to
be started.

- **Graph layout:** GRAB uses a modified version of the Sugiyama layout algorithm (see section 4.1.4) to display the graph.

- **Graphical abstraction:** One of the features of GRAB is that it
  can scale the graph to fit the size of the editing window. The user
  can then "zoom in" or "zoom out" to increase or decrease the scaling
  factor. Assuming that the window size remains fixed, "zooming in"
  may cause part of the graph to no longer be visible. In that case, the
  user can use the scrollbars to view these portions of the graph.

- **Persistence:** The input/output format is a simple list of nodes and
  a list of edges. According to [Mes89], a representation of the graph
  can also be generated for inclusion in troff text formatter documents.
  No explicit support for persistence is available.

- **Extendibility:** There does not appear to be any explicit application
  interface. Apparently, the source code is simply modified to adapt
  GRAB to the new application.

DRAG [Tri88] is a graph drawing program whose goal is to produce high quality layouts suitable for inclusion in publications. Time is not nearly as
important as the quality of the layout (it lays out a 50-node graph in
several minutes on a Sun 3 workstation).

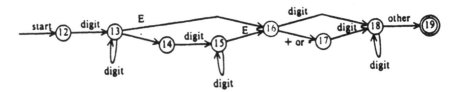

Figure 3.8: DRAG graph drawing program

- **Graph layout:** DRAG offers two layout algorithms. The first is its
  own called the ranked planar layout algorithm and the second is the
  orthogonal minimum bend layout algorithm of [BNT86]. The ranked
  planar layout algorithm first determines the assignment of nodes to
  levels. The level assignment is either given by the user or is determined by a topological sort of the directed graph. Subsequently a
  planar embedding of the graph is determined. The planar embedding
  phase is complicated by the fact that it attempts to take the edge
  routings into account by allocating routing channels around other
  nodes and edges if necessary.

- **Graphical abstraction:** None.

- **Persistence:** The input ranges from a minimal specification describing only the structure of the graph to a more detailed specification in which many display attributes can be specified. One can specify the graph's orientation (left-to-right or top-to-bottom), anchor points where edges may connect to nodes, labels for edges and how they are to be drawn (e.g. dashed, dotted, etc.). DRAG generates output in IDEAL [vW82] format, a high-level language for specifying pictures.

- **Extendibility:** None.

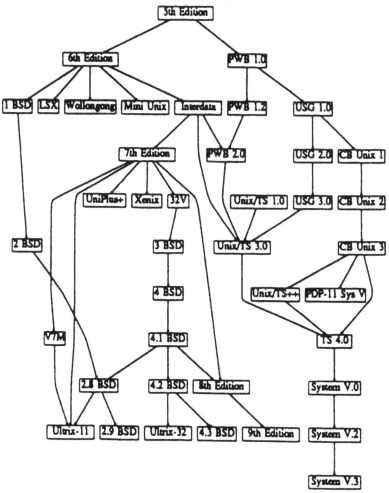

Figure 3.9: DAG graph drawing program

DAG [GNV88] is a graph drawing program which can generate PIC[Ker82] or PostScript[Sys85]. When generating PIC it can fit in the troff-pipeline (dag | pic | troff | lpr). DAG's successor is a graph drawing program called DOT[KN91]. In DOT, nodes can be assigned "ports" and this feature is used to produce "nice" drawings of data structures (e.g. records or hash tables). Additional features of DOT are that edge labels are placed so that they do not overlap parts of the graph's display, an attributed graph description language is used to describe the graph, parallel edges can be merged together, and the layout algorithms have been improved[GKNV].

- **Graph layout:** The layout algorithm is split into phases which roughly correspond to that of the Sugiyama layout (see section 4.1.4). The subproblems themselves are solved using different techniques. The four-pass algorithm [GKNV] is as follows.

  In the first pass, a network simplex algorithm is used to determine the optimal rank assignment of the nodes. One can control the assignment to ranks somewhat by specifying a node should be at the minimum or maximum rank or at the same rank as some other node(s). In the second pass, the relative positions of the nodes within each level are determined by the "weighted median" heuristic together with local transpositions which reduce the number of crossings. The third "finetuning" phase determines the final coordinates of the nodes by constructing and ranking an auxiliary graph. The final pass determines spline points for the edges and places edge labels so that they do not overlap with other parts of the graph.

- **Graphical abstraction:** The user can choose to merge parallel edges together to produce tidier layouts.

- **Persistence:** The input format is an attributed list of nodes and edges. The attributes for node include: shape, size, label, pointsize, and color. The attributes for edge include: label, pointsize, weight, and ink style (solid, dashed, dotted, or invisible). One can lay out the graph top-to-bottom or left-to-right. One can control the x- and y-coordinate spacing between the nodes.

- **Extendibility:** None.

3.3   Extendible Graph Editors

ISI Grapher [Rob87] is a graph browser implemented in Common Lisp. It interfaces easily with other Common Lisp system tools and applications. It has been adapted to the following set of applications: the flavor grapher (displays interdependencies between Lisp "flavors"), the package grapher

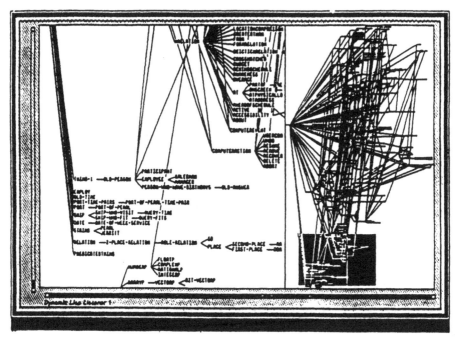

Figure 3.10: ISI editor (from [Mes89])

(displays interdependencies between a Lisp package and all packages that use it), the divisors grapher (displays the divisibility graph of a given integer), and a browser for concept taxonomies. The ISI grapher has a novel way of generating a hardcopy of the graph. A graph is typically too large to fit on a single screen. The ISI grapher automatically incrementally scrolls the graph and invokes the system-dependent dump-to-printer command for each screen full. The user can then cut and paste the pages together to obtain a hardcopy of the entire graph.

- **Graph layout:** It provides a simple automatic layout using a depth-first search algorithm which leads to an efficient (linear) running time. The algorithm can handle a large number of nodes (on the order of 20,000 nodes), and can lay them out quickly (ca. 2500 nodes per minute on a Symbolics 3600 workstation). However, the layout (see figure 3.10) is not pleasing for graphs that are not tree-like. The algorithm places leaf nodes a fixed distance from previously laid out nodes and places other nodes at the average position of their successors. Cycles, which are detected during a topological sort, are handled by breaking the cycle and repeating one of the nodes.

- **Graphical abstraction:** None.

- **Persistence:** The input to the editor consists of a start node and a successor function. There appear to be no facilities for saving the graph.

- **Extendibility:** The ISI grapher provides a set of standard operations to draw a node, highlight a node etc. Each of the standard operations is associated with a "function precedence list". Each operation has an initial default function that returns a non-nil value. The application developer may prepend additional functions to a list. When a standard operation is invoked, the most recently added function in the function precedence list will be invoked. If it returns nil (i.e. failure) then the next function in the list will be invoked until one of the functions returns non-nil (i.e. success).

GraphEd [Him88] is a graph editor intended for viewing and animating graphs. One of the novel aspects of GraphEd is its support for interactive manipulation of graph grammars[ENRE87].

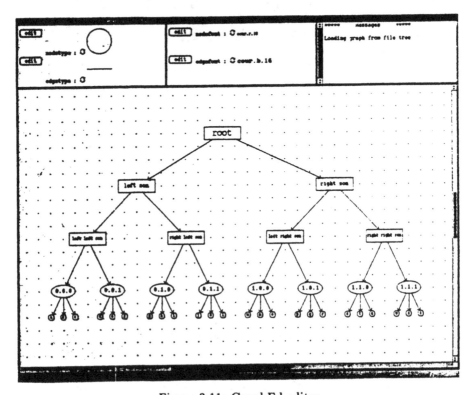

Figure 3.11: GraphEd editor

- **Graph layout:** The user can choose between manual and automatic layout. Currently a version of the Sugiyama layout (see section 4.1.4), a tree layout, and an undirected planar layout are provided. Edges may attach to nodes at the corners of the node or in the center of the node. One of the novel features of this editor is that the edge drawing routine automatically selects the best anchor point. Unlike many other systems, the drawing of multiple edges between nodes is supported.

- **Graphical abstraction:** Cut and paste operations, familiar to users of interactive drawing programs, are also available to copy entire subgraphs. The user can select a set of nodes and perform an action on the entire set.

- **Persistence:** The minimal input to the editor is an adjacency list giving the node's positions. Other information that may be optionally specified in the input includes additional information on the nodes and edges (size, color, font etc.), information about the appearance of node and edge types, and the current size and position of the editing window. The user can define node types and can customize the appearance of the nodes by specifying an icon associated with the node. The appearance of edges can be specified similarly.

- **Extendibility:** GraphEd can be used as a library by various application programs.

Kb-edit [TW87] is an extendible graph editor based on the semantic net formalism. Kb-edit has been adapted to the following set of applications: a PERT chart editor, an editor for graphical software configuration management, a graphical design tool for software, an editor for taxonomies, an editor for graphical process flow models, and a graphical knowledge acquisition tool. Kb-edit is implemented using CRL, an extension of Common Lisp which is part of Knowledge Craft [PK86], an expert system development shell.

- **Graph layout:** All relations have a preferred direction (above, below, left-of, right-of) and automatic layout is performed by displaying nodes such that all other nodes above and to the left have already been displayed. Kb-edit provides one of the rare layout algorithms that takes different preferred edge directions into account.

- **Graphical abstraction:** The user can restrict the displayed portion of the graph using a parameter specifying the maximum path length displayed by the editor. Nodes whose edges are thus "cut off" are specially marked. The user can selectively "expand" such a nodes and can later retract this expansion. Kb-edit also supports multiple views of the graph. A layout overview of the entire graph can be shown indicating the user's current position in the graph.

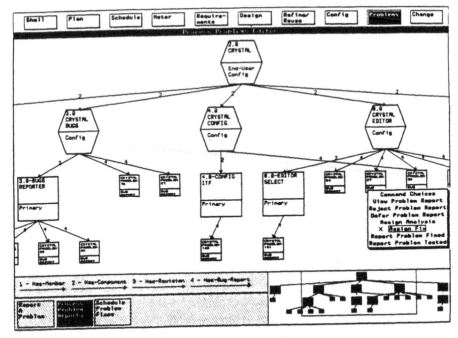

Figure 3.12: Kb-edit editor

- **Persistence:** Persistence is achieved through the underlying Lisp
  system.

- **Extendibility:** The application developer can customize the appearance of the editor by defining a set of node and edge types and their
  appearance. The inheritance mechanism of CRL can be used to simplify the customization. Adding the graph editor user interface to a
  CRL-based application is simply a matter of extending menu entries
  or setting pointer values to functions that are invoked by the graph
  editor.

GMB (Graph Manager/Browser) [JG89] is a graph editor consisting of two
major components – the manager which implements the abstract data type
for the graph and the browser which displays a graph. GMB applications
include a graphical program browser, a graphical version of the make[Fel79]
program, and a dynamic parallel program monitoring tool.

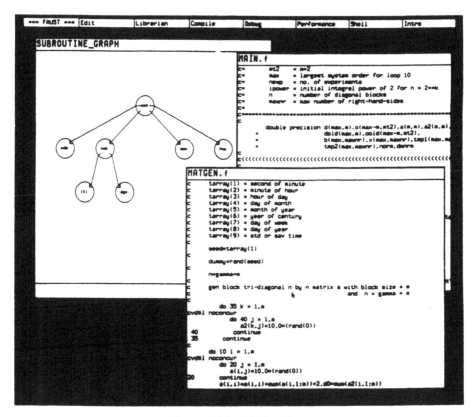

Figure 3.13: GMB editor

- **Graph layout:** GMB offers three alternative strategies for automatic layout: a standard automatic layout algorithm based on the Sugiyama algorithm (see section 4.1.4), a layout provided by the application, or a layout provided by GMB which attempts to combine a Sugiyama layout with simple layout constraints. Most graph editors have a fixed set of anchor points where an edge may connect to a node. One of the novel aspects of the GMB graph layout algorithm is that the choice of anchor point is determined automatically when the edge is laid out. The anchor points are chosen such that, if a node has several incoming edges, their arrow heads do not overlap.

- **Graphical abstraction:** GMB offers a recursive abstraction mechanism called a "node group" which allows a group of nodes to be treated and viewed as a single node. When a new node group is formed, edges are automatically created that show the connection between the node group and nodes outside the group. The user, however, does not have the ability to see the set of nodes in the node group – only the single node representing the group is shown.

The user can group a set of nodes and edges which are of particular interest into a "graph view". The graph view can be displayed in a separate window. Furthermore, several different views of a graph can be shown simultaneously.

- **Persistence:** Graph data structures can be stored in files, but details of the format are not available.

- **Extendibility:** GMB has a complex application interface. For example, in order to display a view of a graph, the application developer must explicitly create the graph, create a view of the graph, specify which nodes and edges should be displayed in the view, and then request that the view be displayed.

GraphView [BS89] is an extendible graphical editing tool for the NeXT computer. The extendibility is achieved by relying on separately-compiled external "transformation" programs. Two special types of transformations are *generators*, which create a new graph, and *displayers*, which perform graph layout. To apply a transformation, an ASCII representation of the graph is saved in a file, the external program is invoked, and thereafter the modified file is read back in. A history of the session is maintained by keeping a list of the transformations performed on the graph.

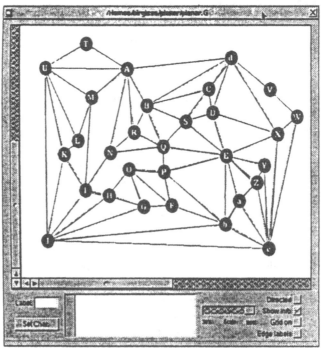

Figure 3.14: GraphView editor

- **Graph layout:** GraphView does not provide any graph layout algorithms, but the application developer can provide a layout algorithm as a transformation.
- **Graphical abstraction:** The user can display several graphs simultaneously in separate windows and the editor allows the user to cut and paste subgraphs between windows. No other abstraction mechanism is provided, although, as with the layout algorithm, this could be accomplished via an external transformation program.
- **Persistence:** A textual representation of the graph can be saved in a file. This representation is used to achieve persistence as well as to communicate with the transformation programs.
- **Extendibility:** This reliance on external transformations allows extendibility without having to restart or recompile the graph editor itself, but of course the reading and writing of the graph will have an impact on the performance of the editor. The implementation language is Objective-C, an object-oriented extension to C.

3.4 Summary

System	Graph Layout	Graphical Abstraction	Persistence	Extendibility
GINCOD	yes	yes	yes	no
IDETool	no	no	yes	some
PROSPEC	no	yes	yes	no
ParaGraph	yes	no	no	no
GERM	no	yes	yes	no
VIFOR	yes	yes	yes	no
GRAB	yes	some	no	no
DRAG	yes	no	no	no
DAG	yes	no	no	no
ISI Grapher	yes	no	no	yes
GraphEd	yes	yes	yes	some
Kb-edit	yes	yes	some	yes
GMB	yes	yes	yes	some
GraphView	no	no	yes	yes

Table 3.1: Comparison of graph editors

Table 3.1 shows a summary of the systems surveyed. Each entry indicates whether the system supports, does not support, or only partially supports the facilities of graph layout, graphical abstraction, persistence, and extendibility.

In the "Motivation and Goals" section of the Introduction, desirable properties of a graph editor were listed. These properties were distilled, in part, by examining existing graph-based tools and identifying problem areas.

Some of the programs surveyed in this chapter are particularly relevant to the design of the extendible graph editor presented in this dissertation. The graph drawing programs DAG and DRAG provide sophisticated layout mechanisms because they take some layout constraints into account. These two programs are also relevant in the area of persistence because they both use an input format which can range from a simple list of nodes and edges to a more complex description of the graph and its display. The graphical abstraction mechanisms available in GERM and GMB underscore the usefulness of graphical abstraction. The navigation aides provided by the ISI Grapher, GERM, and Kb-edit demonstrate their practicality. GMB and GraphView show that object-oriented design is an appropriate mechanism to achieve extendibility.

None of the systems surveyed provide adequate solutions to all of these problem areas. The following four chapters will present research in each of these problem areas and propose a solution to each.

Chapter 4

> *"Without the requirement*
> *of mathematical aesthetics*
> *a great many discoveries*
> *would not have been made."*
> *– A. Einstein*

Layout Algorithms and Layout Constraints

This chapter presents a layout constraint manager which can be used together with existing graph layout algorithms. This technique can be used, not only to take user- and application-specific layout constraints into account, but also to achieve layout stability[BP90].

4.1 Related Work

A *layout algorithm* determines the placement of each node in a graph. The routing of each edge is often determined as well. The use of a layout algorithm relieves the user of the tedious and error-prone manual placement of the nodes and is thus especially useful for large graphs. Generally, the desired end result is to produce a drawing of the graph that is easy for the user to understand. This is accomplished by trying to meet a set of aesthetic goals. A *layout constraint* is a restriction placed on the layout. Typically, the restriction involves the placement of one or more nodes of the graph, for example that one node should be placed to the left of another node or that a set of nodes should be placed together.

An excellent overview of graph layout algorithms can be found in [TBB89]. There, graphs are categorized according to several parameters including the graphic standard, the aesthetics, and the layout constraints. The *graphic standard* determines whether the nodes are placed according to the grid or straight-line standard. The *grid standard* forces the nodes to be placed on a grid. Edges are often routed along the grid as well. The *straight-line standard* allows arbitrary placement of the nodes, and edges are drawn as straight lines between the nodes. These two standards may also be mixed, for example, allowing arbitrary placement of the nodes and splines or a set of edge segments for the edges.

The *aesthetics* of a graph measure how "good" the appearance of the graph is. The following is a partial list of aesthetics discussed in the literature.

Minimize edge crossings	Place nodes and edges such that the number of edge crossings is minimized.
Hierarchical	Maximize the number of edges that point in the same general direction.
Maximize symmetry	Display symmetry inherent in the drawing.
Minimize area	The total area used by the graph layout should be minimized.
Minimize width	The width of the total area used by the graph layout should be minimized.
Minimize longest edge	The length of the longest edge is minimized.
Minimize total edge length	Place nodes and edges such that the total edge length is minimized.
Uniform edge length	Place nodes and edges such that all edges are of approximately equal length.
Uniform node distribution	Place nodes such that the spacing between nodes is approximately equal.
Minimize edge bends	Place nodes and edges such that the number of edge bends is minimized.
Aligned levels	Place nodes of equal depth at the same level.
Equi-spaced levels	Leave same amount of space between each pair of levels.
Centered parents	Center parent nodes between their children.
Centered children	Center nodes between their parents.
Isomorphic subgraphs	Subgraphs with the same structure should always be displayed alike.

Several layout algorithms try to satisfy more than one aesthetic simultaneously. Trying to satisfy more than one aesthetic simultaneously can, of course, lead to conflicts. For example, insisting on a hierarchical layout may cause edge crossings that could have been avoided in a planar layout, as shown in figure 4.1.

The choice of aesthetics depends on the type of graph, the application, the user's taste, and possibly other factors. For example, if the graph's edges represent precedence relations, then clearly it is important to display the hierarchy. Similarly, if the graph is undirected, then taking advantage of symmetry and displaying isomorphic subgraphs alike is important. Speed will be much more important for an interactive graph browser for large graphs than for a graph drawing program which produces proof-ready pictures of small graphs for typesetting.

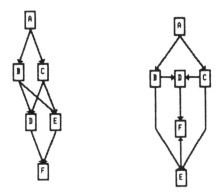

Figure 4.1: Planar layout can eliminate crossings

Tamassia[TBB89] limits the definition of *layout constraints* to restrictions that are placed on the layout based on the semantics of the graph. The definition used here is more general, allowing restrictions to be placed on the layout of the graph for any reason.

Some examples of layout constraints from the literature include:

Straight line	A group of nodes should be aligned.
2-1/2D	A group of nodes should be displayed so that they overlap one another.
Cluster	Keep a group of nodes near each other.
Center	Place a particular node or group of nodes in the center of the graph.
Boundary	Place a particular node or group of nodes at the outer edges of the graph.
Relative	Place a node above/below/left-of/right-of another node.

As with aesthetics, if more than one type of constraint should be satisfied it is desirable to rank the relative importance of the constraints. In some cases it may be reasonable to include the aesthetic goals in this ranking rather than to satisfy the constraint at any cost. The problem of determining the layout that best satisfies the constraints can be difficult and time consuming.

The layout aesthetics used by VLSI layout algorithms are quite different from those used in a graph editor and therefore they are not relevant here. For example, some VLSI layout algorithms aesthetics are efficient use of space, grouping a set of edges together in a "channel", or minimizing the number of layers required. Furthermore, users of VLSI layout algorithms put much more emphasis on producing a good layout than producing a layout quickly.

The remainder of section 4.1 presents several layout algorithms grouped according to the type of graph. The annotated bibliography [ET89] is an excellent source for further information on a wide variety of graph layout algorithms.

 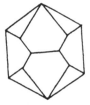

Figure 4.2: Planar drawings: straight line, and convex (from [CON85])

4.1.1 Planar Graphs

Planar layouts are important for graph theory applications, and for various applications where edges have dual directions such as communication networks or traffic routes. The most obvious aesthetic goal is to produce a layout with no edge crossings.

Usually the first step of any planar layout algorithm is to determine whether the graph is planar. Planarity testing algorithms are given in [HT74, BL76, Tut63]. If a graph is "almost" planar, then one can usually still apply a planar layout algorithm by treating each crossing as an additional "crossing" node.

There are several categories of planar graph layout algorithms:

- A *straight line drawing* places the nodes and edges such that each edge is a straight line. A proof by Fary [Far48] states that any planar graph can be drawn using a straight line drawing.

- A *convex drawing* places nodes and edges such that each face boundary is a convex polygon. This often produces a pleasing drawing of the graph, but not all planar graphs can be embedded in a convex drawing. This method was first investigated by Tutte [Tut63]. The algorithm in [CON85] produces a convex drawing in linear time. Figure 4.2 shows two views of a planar graph: one with straight lines and one as a convex drawing.

- In the *grid representation* , edges are connected sequences of edge segments. Often, the further restriction is made that all edge segments be either horizontal or vertical. With this layout, minimizing the number of edge bends is usually one of the aesthetic goals. In [Tam87], Tamassia gives an algorithm that finds the minimum number of bends in $O(n^2)$.

- A *visibility diagram* [OvW78] is a planar embedding where nodes are represented as horizontal segments and edges are represented as vertical segments. The edge segment (a,b) has its endpoints on the vertex segments associated with a and b. In [TT89] Tamassia and Tollis present an algorithm for constructing visibility diagrams in linear time. Figure 4.3 shows a grid and visibility diagram representation of a graph.

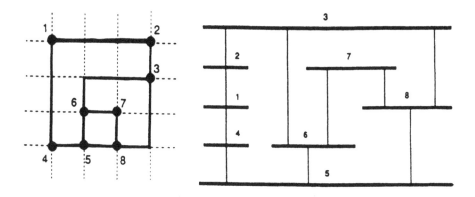

Figure 4.3: Planar drawings: grid and visibility diagram (from [ET89])

- Woods [Woo81] presents a basic algorithm that produces a planar layout in linear time with non-linear-time enhancements to further improve
  the layout of the graph. The algorithm has three phases. First, a planar embedding of the graph is determined using the planarization testing
  algorithm by [HT74]. The second phase assigns the nodes to levels and determines the ordering within each level. A technique called s-t numbering
  is used to number the nodes in each biconnected component of the graph
  so that all nodes except those on the top and bottom levels are connected
  to both nodes above and below them. For biconnected components, such a
  numbering is guaranteed to exist [Eve79]. Finally, the actual coordinates
  of the nodes are calculated.

- Symmetry is an important aesthetic when displaying planar graphs. Efficient algorithms for displaying trees, outerplanar graphs and embedded
  graphs such that the number of symmetries is maximized are presented
  in [Man90b].

4.1.2   Undirected Graphs

An example of this type of graph is an entity-relationship diagram. The aesthetic
goals for undirected graphs are quite general because there is little information
to base a layout on. They include: minimize the area used by the graph, maximize symmetry, minimize edge crossings, minimize edge bends, uniform node
distribution, and uniform edge distribution.

The algorithm presented in [LNS85] tries to maximize the symmetries, but
this involves solving the apparently intractable problem of computing the au-
tomorphism group of a graph. An entity-relationship diagram layout is given
in [TBT83]. The spring embedder algorithm [Ead84, KK89, FR91] represents

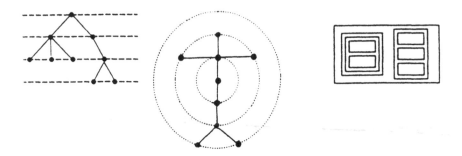

Figure 4.4: Tree drawings: conventional, radial, contour (from [ET89])

nodes as steel rings and edges as springs and calculates a layout such that the energy of the springs is minimized. This produces a natural distribution of the nodes.

A common approach in laying out undirected graphs, is to first make the graph planar by adding "dummy" nodes at the crossing points and subsequently applying a standard planar graph layout algorithm.

4.1.3 Trees

Examples of where tree layouts are appropriate include syntax trees and decision trees. The aesthetic goals for layout algorithms for trees are fairly well established: hierarchical, aligned levels, equi-spaced levels, and centered parents. Edges are usually drawn as straight lines between the nodes, but are sometimes drawn along a horizontal/vertical grid instead.

One of the earliest tree layout algorithms is due to Knuth [Knu71]. A recursive equivalent due to Wirth is presented in [Wir76]. Wetherell and Shannon [WS79] show that this algorithm can produce drawings that are much too wide, and they provide an algorithm that meets the additional aesthetic goal of minimizing the width of the layout. Reingold and Tilford [RT81] address the problem of drawing isomorphic subtrees alike each time and show that sometimes it is not possible to meet both the isomorphic subtree and the minimum width goals simultaneously. The algorithm by Manning [MA86] produces an alternative representation by taking advantage of radial as well as axial symmetries. This is referred to as a spiral or radial layout. A further alternative is to show the tree as a nested set of contours where a child's contour is embedded within the parent's contour. Figure 4.4 show a conventional, radial, and contour layout of a tree.

4.1.4   Directed Graphs

Examples of directed graphs include PERT charts, call graphs, and finite state automata. An important aesthetic goal for most directed graphs is for the direction of the edges to be obvious without having to always search for the arrow head. Usually this means having all edges point in the same general direction, but a radial layout might also be appropriate. As with trees, a common aesthetic goal for directed graphs is to align nodes of the same rank. Additional aesthetic goals are to minimize the area and minimize the edge crossings. Minimizing the edge crossings for directed acyclic graphs, however, is an NP-Complete problem [Joh82] even for the simple case of a bipartite graph.

The article [ES90] presents a general method for laying out directed graphs in which the aesthetic criteria are viewed as goals of optimization problems. Several specific layout algorithms, including Carpano [Car80], Sugiyama et al [STT81], Rowe et al [RDM+87], Reggiani and Marchetti [RM88] and Gansner et al [GNV88], are compared.

One of the difficulties with directed graphs is that they can contain cycles. Two common techniques for handling cycles in directed graphs are
   1) group all nodes in the cycle together into a "proxy" node representing the cycle, or
   2) temporarily reverse the direction of the cycle-causing edges.
The algorithm then lays out the resulting directed acyclic graph. Finding the minimum subset of edges to be reversed is an NP-Complete problem called the "minimum feedback edge set" [ET89].

In [BT88], an efficient algorithm for constructing planar upward drawings of graphs is presented.

The ISI Grapher [Rob87] is based on a fast (linear time) but trivial layout algorithm for directed graphs. The algorithm places leaf nodes a fixed distance from previously laid out nodes and places other nodes at the average position of their successors. This layout works best for tree-like graphs.

Warfield [War77] provided the basis for several layout algorithms [Car80, STT81] by suggesting that a directed graph be made into a *proper hierarchy* (i.e. where every edge is between adjacent levels) and that the crossings between adjacent levels be minimized. Minimizing the number of crossings between adjacent levels is shown to be NP-hard in [GJ83].

In [STT81], Sugiyama et al present an algorithm which uses a heuristic to reduce the number of edge crossings. This algorithm, referred to henceforth as the *Sugiyama layout algorithm*, consists of three phases:

- *Preprocessing:* Assign levels to all nodes by sorting them topologically. Eliminate cycles by grouping all nodes in a cycle into a "proxy" node. For "long" edges (those that cross several levels), introduce invisible dummy nodes at all intermediate levels to guarantee that each edge crosses only one level.

- *Barycentric ordering:* Rearrange the nodes on each level with the goal of reducing edge crossings between adjacent levels. Initially, assume that the ordering of nodes in the top level is fixed. Each node in the next level is positioned based on its *down-barycenter*, which is the average position of its predecessors. Continue this downward pass until the bottom level is reached and then do a similar upward pass based on the *up-barycenter*, the average position of the successors. Perform several upward and downward passes through the graph until no more improvement occurs or a user-specified threshold for the number of passes is reached. This is the most computation intensive part of the layout algorithm.

- *Finetuning:* Determine final x- and y-coordinates for each node such that the total edge length is minimized. The y-coordinates are set to the product of the vertical spacing factor and the level number of the node. As in the barycentric ordering phase, perform several upward and downward passes through the graph to determine the x-coordinates. Initially, set the x-coordinates of the nodes in the top level so that they are spread apart by the horizontal spacing factor. Determine the order in which nodes in the next level will be placed – dummy nodes will be placed first followed by nodes with the largest number of incoming and outgoing edges. Calculate the desired x-coordinate of each node as the average position of its predecessors or successors. Place each node as close to its desired position as possible without shifting previously placed nodes or changing the relative order of the nodes. A dummy node will always have exactly one predecessor and one successor, so it will always request to be placed at the same x-coordinate as its predecessor or successor. Dummy nodes are positioned first to align them and thus straighten "long" edges.

This basic algorithm has been extended and improved upon by several developers of tools involving graph layout.

- In the implementation used by the EDGE graph editor (see chapter 8), the dummy nodes in the middle of a level are given higher priority than those at the extremities. Otherwise the finetuning phase may cause a graph to appear unbalanced.

- Davis [Dav86] changed the way cycles were handled. As soon as a cycle is detected, the edge causing the cycle is temporarily reversed. This usually provides a more readable layout than using a "proxy" node, but the results depend on which edge is chosen for reversal. Davis suggests a heuristic of placing nodes in the cycle such that nodes of high in-degree are assigned low level numbers and nodes of high out-degree are assigned high level numbers.

- Davis and Rowe [Dav86, RDM⁺87] modified the calculation of the barycen-
  ters so that after a single downward (using predecessor positions) and a
  single upward pass (using successor positions), further passes were made
  using the average positions of both the predecessors and successors. For
  graphs whose down-barycenters produced a markedly different result than
  the up-barycenters, this modification led to a consensus more quickly.

- Eades [ES90] suggests using a heuristic based on the median as opposed
  to the barycenter (average) to determine the relative positions of nodes
  within a level. One advantage of the median heuristic is that, given a
  bipartite graph, the number of crossings can be shown to be within a
  factor of three of the optimal [EW89]. In [GNV88, GKNV] a version of
  the median heuristic is also used.

- Another of the disadvantages of the Sugiyama layout algorithm is that the
  topological sort does not necessarily lead to an optimal level assignment.
  Figure 4.5 shows how a good level assignment can reduce the total edge
  length. Sometimes edge crossings can be avoided by additional levels in the
  graph, as shown in Figure 4.6. In [Sug84], Sugiyama suggests a heuristic
  for making better level assignments. In [GNV88, GKNV], an optimal
  level assignment is determined by a combinatorial variation of the simplex
  method.

- Because such layouts only optimize locally, they cannot eliminate some
  edge crossings that arise from global situations. Davis [Dav86] suggests
  using a barycenter value that is weighted such that it not only considers
  the adjacent level, but also (with lesser weight) more distant levels.

- In [MRH89] a "divide and conquer" approach to graph layout is presented.
  Here it is shown that for large graphs, one can often produce faster and
  better layouts by partitioning the graph into subgraphs, laying out each
  subgraph separately using a directed graph layout algorithm based on the
  Sugiyama approach, and then merging the subgraph layouts back into one.

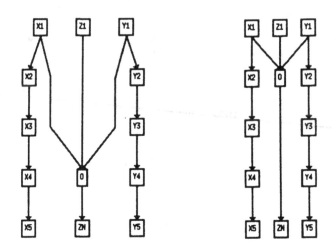

Figure 4.5: Level assignment can affect total edge length

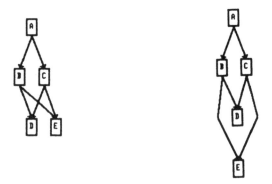

Figure 4.6: Level assignment can affect number of crossings

4.2 Layout Constraints

As shown in the previous section, graph layout algorithms place nodes and edges of a graph based on a set of aesthetic goals. Typically, the layout algorithm has no knowledge of the semantics of the graph and simply produces a readable and comprehensible drawing of the graph. However, in many cases it is desirable that the user and/or the application be able to specify constraints on the layout of the graph. This section discusses what type of graph layout constraints are appropriate and describes a general mechanism for extending graph layout algorithms to be able to handle constraints.

The earliest use of constraints for graphical layout was in Sketchpad [Sut63]. In ThingLab [Bor79], a constraint-based simulation laboratory, many of these ideas were extended and realized in a Smalltalk environment. The Juno [Nel85] system combines a constraint-based language for describing pictures with a WYSIWYG picture editor. In contrast to previous systems where constraints were always explicitly specified by the user, the Peridot system [Mye87, Mye90] infers simple graphic constraints automatically. Recently, constraints have been applied to user interface toolkits, where constraints are combined with "active values" to bind graphical objects to the application's data structures (see [SM88, MBFB89]).

A spectrum of algorithms for solving constraint hierarchies is presented in [FBMB90]. The spectrum represents the trade-off between generality (red) and efficiency (blue). An incremental version of the fast local-propagation ("blue") algorithm called DeltaBlue is presented. This algorithm is currently used in ThingLab II [MBFB89], a descendant of ThingLab, where it is used for interactive user interface construction. For more information on constraints and constraint programming languages in general, see [Lel88].

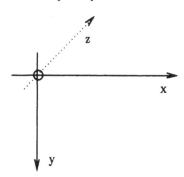

Figure 4.7: Coordinate system

Figure 4.7 shows the coordinate system which will be used throughout. The x-axis runs from left to right, the y-axis from top to bottom. The origin of the coordinate system is assumed to be in the upper left corner. For three-

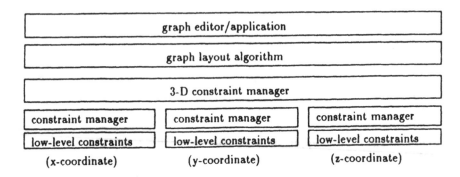

Figure 4.8: Overview of constraint manager architecture

dimensional layout there may also be a z-axis running from the back to the front.

The evaluation of a set of n single-dimensional constraints takes $O(n^3)$ operations [Dav87]. However, if the constraints are further restricted to being of the form $a = b + c$ or $a \geq b + c$ then a run time of $O(n)$ is attainable. In this case, a constraint propagation algorithm based on a topological sort [Böh89] can be performed in linear time.

How much of a restriction does this place on the list of available constraints? The major restriction here is that constraints can only be given in a single dimension. Constraints such as $A_x = A_y$ (i.e. A should be placed on the diagonal) cannot be expressed. Constraints can, however, be expressed as a set of several independent single-dimensional constraints. For example, the request that "node A should be placed vertically above node B" can be expressed with the two independent constraints $A_x = B_x$ and $A_y < B_y$. The sort of constraints likely to be used by a graph editor are relatively simple. In fact, many of them can be expressed as constraints in a single dimension – i.e. place node A to the left of node B.

An overview of the proposed s...ution is shown in Figure 4.8. Constraints for a single dimension are specified at the lowest level. The next higher layer manages those constraints. The 3-D constraint manager combines the constraints from the three dimensions. The next layer is the layout algorithm. The topmost layer is the graph editor and the application. A strict interface is used between the constraint manager and the graph layout algorithm to keep them independent.

The following subsections describe each of these layers in more detail. One particular layout algorithm, the Sugiyama layout algorithm, is used throughout as an example, but because of the the strong separation between the 3-D constraint manager and the layout algorithms, the same approach would work for many other layout algorithms as well.

4.2.1 Low-level Constraints

A *low-level* (single-dimensional) *constraint* is defined as a linear equation of two variables. A set of these constraints is called a *constraint network* . The constraints can be formulated using the coordinates of each node (e.g. $A_x = B_x$). Each constraint may also be given an integer priority value indicating the importance of satisfying this constraint, and the set of constaints is thus said to form a *constraint hierarchy*. To constrain the position of a node in a graph, there are three different types of constraints:

Absolute positioning: Constrain the node's position in relation to a fixed coordinate system. For example, assuming that nodes are placed in horizontal levels, constrain placement of a node to a particular level ("level 4") or to a range of levels ("levels 3–5").

Relative positioning: Constrain the node's position in relation to other nodes. For example, "node A is to the left of node B" or "node C is the neighbor above node D".

Clusters: Gather a group of nodes together to a cluster which can then be further constrained. For example, "put nodes A,B,C into a cluster named E", "cluster E must have a maximum width of 3 units", or "place all nodes in cluster E to the right of node G".

The following single-dimensional constraints can be supported by a constraint manager based on linear equations:

- *equal, smaller, greater*: place a node at the same, smaller, or greater position than another
- *neighbors*: place nodes adjacent to another
- *low_margin, high_margin*: place the node at the absolute smallest or largest position
- *range*: specify a range of absolute positions a node may occupy
- *cluster*: group a set of nodes together in a cluster
- *limit*: limit the number of positions within a cluster

4.2.2 Constraint Manager

For each dimension, there is a constraint manager which has two main tasks:

- Maintain a list of all constraints and provide functions to add, delete, and check the status of constraints in the constraint network.

- Evaluate the constraint network and keep it consistent. A set of constraints is defined to be *consistent* if none of them are contradictory.

The purpose of the evaluation of the constraint network is to compute the global effects of local constraints. For example, from the chain of order relations $A_x < B_x$, $B_x < C_x$, $C_x < D_x$, the relation $A_x < D_x$ should be derived. This can be done in linear time in the number of constraints by a technique called "constraint propagation" which is similar to topological sorting. After this preprocessing step, queries can be answered in constant time. This efficiency is extremely important for layout algorithms which may make extensive computations while reordering nodes in the graph layout.

In the case where the set of constraints is not consistent, the constraint manager computes a subset of consistent constraints by deactivating some of the constraints. The selection of constraints to be deactivated is influenced by the priority value associated with each constraint. The constraint manager tries to keep high priority constraints active by deactivating low priority constraints first. Among inconsistent constraints with equal priorities the selection is arbitrary. Deactivated constraints are ignored during the evaluation of the constraint network.

The detection of which constraints are deactivated is done by a binary search through inconsistent sets of constraints until all constraints causing an inconsistency are deactivated. The set of constraints is maintained on a list which is sorted by priority. When an inconsistency is detected, the list is split in half and the constraints in the high-priority half are tested for consistency. This process is repeated until a consistent set of constraints is determined. This approach does not necessarily lead to the maximum subset of consistent constraints, but does produce acceptable results. The performance of the consistency checker is crucial. If a set of constraints is consistent, the check runs in linear time in the number of constraints. However, finding and removing constraints that cause inconsistencies can slow down the algorithm noticeably ($O(n^2 \log n)$ in the worst case for n constraints[Böh89]). Although a more efficient solution for handling inconsistencies would be beneficial, experience has shown that the number of inconsistencies in practical cases is much smaller than the number of constraints.

4.2.3 Three-dimensional Constraints

So far the dimensions have been treated independently. However, in order to define a convenient interface to the user, layout algorithms, and applications programs, an interfacing module is used. This module provides functions for each of the aforementioned single-dimensional constraints. Additionally, this module provides several procedures that translate commonly used two- or three-dimensional constraints into several (independent) single-dimensional constraints.

4.2.4   Integration with Layout Algorithms

This section shows how layout constraints may be integrated into a layout algorithm. The following describes the integration into Sugiyama layout algorithm, but the same methodology can be applied to different layout algorithms as well.

In order to understand the extent to which the Sugiyama layout algorithm was modified, review the description of the algorithm in section 4.1.4. For each of the phases some changes or extensions to the original algorithm were necessary. The correspondence between the coordinate system used by the constraints and the layout algorithm is as follows. In the x-direction, coordinate units correspond to subsequent positions. In the y-direction the levels are assigned subsequent numbers. So the constraint "A is the left neighbor of B" can be defined by the two equations "$A_x + 1 = B_x$" and "$A_y = B_y$". The following text assumes that the orientation of the graph is top-to-bottom. Naturally, the same procedure applies if a different orientation (e.g. left-to-right) is used.

- *Preprocessing:* Instead of performing a topological sort, a constraint of the form $smaller(A_y, B_y)$ is generated for each edge from A to B. Evaluation of these constraints (together with any user- or application-specific constraints) yields a valid level assignment. In contrast to the Sugiyama layout, constraints make it possible that the source and target nodes of an edge be on the same level if the user or application so requests. For each such "flat" edge, the additional constraint $neighbors(A_x, B_x)$ is generated so that the source and target nodes will be adjacent on the level. Similarly, for nodes constrained to lie in front of each other, additional constraints are generated which request that these nodes be placed as neighbors on the same level. For "long" edges between nodes that are constrained to be vertically aligned, additional constraints are generated to request that all intermediate dummy nodes also have the same x-coordinate.

- *Barycentric ordering:* The main work of the Sugiyama layout is done during the barycentric ordering phase when the ordering of the nodes within each level is rearranged iteratively according to the average position of its successors/predecessors. The total ordering determined by the Sugiyama layout algorithm is translated into a set of constraints on the x-coordinates of the nodes. For example, $A \rightarrow B \rightarrow C$ is translated into the constraints $smaller(A_x, B_x)$, $smaller(B_x, C_x)$, and $smaller(A_x, C_x)$. If the entire set of constraints is consistent, then the barycentric ordering remains intact. Otherwise, the constraint manager will deactivate constraints to determine a consistent subset of constraints. The layout algorithm will query the constraint manager each time it wants to move a node. For this reason, it is important that the query of the constraint manager be as efficient as possible.

- *Finetuning:* The only change for the finetuning phase is that nodes that differ only in the z-coordinate (depth) are placed such that they overlap slightly (see figure 4.12(c)).

This same technique could be used to extend other layout algorithms as well. The only prerequisite is that the layout algorithm has steps where it is reasonable to insert a query to the constraint manager concerning the placement of nodes. For example, consider a simple tree layout algorithm. Such an algorithm normally places the children of a node on the next level so that they are centered under the parent, but the relative position of the children is irrelevant. Such a tree layout algorithm could be extended to query the constraint manager prior to placing the children. For example, in a tree layout representing parts of a compiler, this approach could be used to place nodes representing the lexical analysis, the syntactic analysis, the code generation, and the code optimization in a natural left-to-right ordering. It would be slightly more difficult, but also possible, to control the (vertical) level assignment of nodes in a tree layout algorithm.

4.2.5   Examples

This section shows three examples of how constraints can be used to make a graph more readable and understandable. Two graphs are shown for each example – the first shows the layout using the Sugiyama algorithm without layout constraints and the second shows the layout with layout constraints taken into account. Figure 4.9 shows a family tree where two kinds of constraints are used. When displaying a family tree, the husband node is placed to the left of the wife node and the children are ordered left-to-right based on their age. Figure 4.10 shows a PERT chart application in which the critical path of the "project" is constrained to appear vertically aligned. Figure 4.11 shows a time line of the development of various UNIX tools[1]. Constraints are used to place a tool according to the year in which it was developed. Additional nodes representing the year of development are added to the graph and are used to constrain the tools to appropriate year.

---

[1] This example is taken from the DOT User's Manual[KN91].

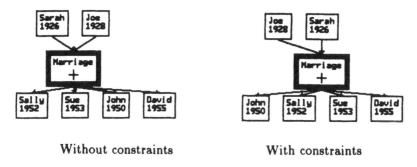

Without constraints   With constraints

Figure 4.9: Constraint example: family tree

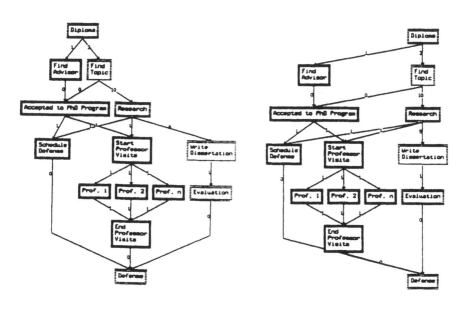

Without constraints   With constraints

Figure 4.10: Constraint example: PERT chart

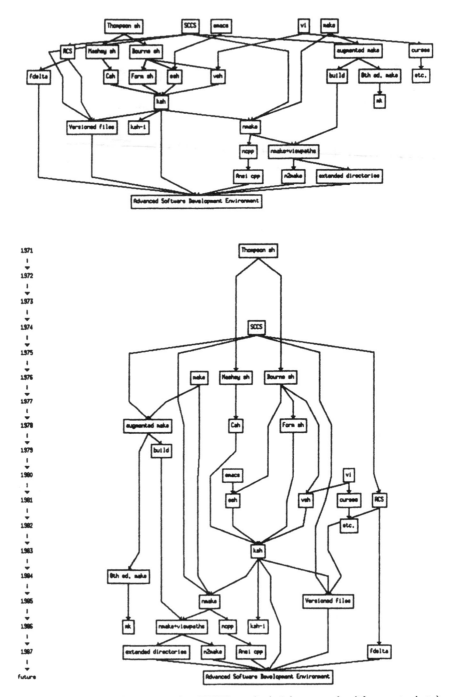

Figure 4.11: Constraint example: UNIX tools (without and with constraints)

| Name of Graph | Layout Time | |
	Without Constraints	With Constraints
Family tree	1 sec.	1 sec.
PERT chart	1 sec.	1 sec.
UNIX tools	4 sec.	10 sec.

Table 4.1: Layout time comparison

4.2.6 Results

The three examples shown above demonstrate that the single-dimensional layout constraints provided by the proposed solution can significantly improve the layout of the graph. Table 4.1 shows the amount of time taken by the layout of each of these graphs, both with and without layout constraints measured on a Sun 3 workstation. From this table, it is clear that the improved layout can often be obtained without significant increase in the layout time.

The limitations of this approach are the following:

- In order to add the layout constraints to a layout algorithm, the layout algorithm has to have apppropriate places to insert the queries to the constraint manager. Some layout algorithms are easier than others to adapt in this respect.

- Because multi-dimensional constraints are split into several single-dimensional constraints which are subsequently treated separately, the situation can arise where one part of a user-specified constraint is satisfied and another part is not. This may be disconcerting to the user.

- The current solution to resolving inconsistencies is simple. When an inconsistency is detected, a relatively large number of constraints may be deactivated to obtain a consistent subset.

The last two items are limitations which were conciously made for efficiency reaons. Other constraint solvers, in particular the incremental DeltaBlue algorithm [FBMB90], are worthy of future investigation.

4.3 Layout Stability

Many layout algorithms do not take the current layout into account when a new layout is requested. This often results in a new layout which looks substantially different from the previous one and this can be disconcerting for the user[2].

---

[2] In fact, the subgraph the user was just editing might no longer be displayed in the scrollable region of the editing window.

Layout stability is a measure of how much a graph changes between successive layouts. Ideally, one would like to minimize the "difference" between successive layouts while still taking the aesthetic goals into account [TBB89].

Although it is generally agreed that layout stability is a serious problem with layout algorithms, layout stability is still a relatively unexplored research area. The solution proposed here uses layout constraints to achieve layout stability. There are four basic steps to this procedure.

- *Generate stability constraints:* The system generates a (rather large) set of constraints which describe the current layout. Constraints are generated to record the level assignment of each node and to record the relative position of nodes within each level. These generated constraints are given a low priority so that they will be deactivated if a conflict with user- or application-specified layout constraint occurs.

- *Determine the vicinity of change:* Any change in the graph's structure can be said to *directly affect* a subset of nodes in the graph. Addition or deletion of a node directly affects that node and addition or deletion of an edge directly affects the source and target nodes of that edge. The *vicinity of change* is a set of nodes that are directly or indirectly affected by the change. Whether a node is indirectly affected is determined from how many edge lengths away the node is from a directly affected node. The *vicinity radius* is a positive integer parameter which specifies the maximum number of edge lengths. A vicinity radius of 0 restricts the vicinity of change to node(s) directly affected by the change. At the other extreme, a large vicinity radius (e.g. equal to the longest path in the graph) can cause all nodes in the graph to be in the vicinity of change.

- *Deactivate constraints in the vicinity of change:* All nodes in the vicinity of change are freed from their stability constraints.

- *Perform a new layout:* A new layout is performed which takes the remaining stability constraints into account as well as any user- or application-specific constraints. Recall that the stability constraints will have lower priority than other constraints, and are thus more likely to be deactivated if inconsistencies are detected. During the new layout, the nodes in the vicinity of change will be free to move whereas the rest of the graph will remain relatively stable. If all nodes on a level remain stable, their relative positions need not be recomputed. This tends to compensate for the additional costs incurred by the stable layout.

The user can specify the vicinity radius and thereby control the extent of the layout stability. Specifying a vicinity radius of 0 forces all nodes to keep their old positions. At the other extreme, specifying a huge vicinity radius allows all nodes to move – thus resulting in the same layout as would be obtained by selecting no stability.

4.3.1 Examples

Figure 4.12 shows the effect of layout stability on a small graph. The graphs
in the top part of figure 4.12 have no user-specified constraints. The graph in
the bottom part of figure 4.12 has the following three constraints: Nodes "C"
and "G" should have the same x-coordinate, nodes "F" and "G" should have
the same y-coordinate, and the nodes "$D1$," "$D2$," and "$D3$" should be shown
in front of each other. Figures 4.12(b) 4.13(b) 4.14(b) show how the graph from
4.12(a) is affected by adding an edge from B to G. Figure 4.12(b) shows the
change without layout stability. Note that the graph has changed considerably.
Figure 4.13(b) shows the change with layout stability and a vicinity radius of
0. Note that, although the graph's layout is worse in terms of edge crossings,
it may be preferable to the user because of its stability. Figure 4.14(b) shows
the change with layout stability and a vicinity radius of 1. Note that this figure
has fewer crossings than 4.13(b); in fact it has the same number of crossings
as the layout in 4.12(b), yet the positions of the nodes remain relatively stable.
Figure 4.12(d) shows how the graph from figure 4.12(c) changed after adding
the edge from B to G. In this case, layout stability was not used and only
user-specified constraints were active. This demonstrates that, in small graphs,
user-specified constraints may be sufficient to achieve stability.

While the previous example is useful for explaining how layout stability
works, it is too small to show the dramatic affect that layout stability can have
on a graph. Figures 4.15 through 4.23 show three further examples. A vicinity
radius of 0 was used for the stable graph layouts in all three of these examples.

Figure 4.15 shows a graph depicting variables used to simulate the future
dynamics of the world (43 nodes, 67 edges). This example originally appears
in [For71], is reprinted and used as an example in [STT81], and is one of
the standard examples used to compare hierarchical layout algorithms (e.g.
see [Mes89, GNV88, RDM+87]). Consider the addition of an edge between
nodes "18" and "6". Figure 4.16 shows the resulting graph when no layout stability is used. Figure 4.17 shows the resulting graph when layout stability is
used.

A second example is shown in the set of graphs in figure 4.18 (41 nodes, 49 edges). Figure 4.18 shows a graph depicting the UNIX development[3]. Consider the addition of an edge from "LSX" to "$7thEdition$". Figure 4.19 shows the resulting graph when no layout stability is used. Figure 4.20 shows the resulting graph when layout stability is used. Note that the nodes "LSX" and "$7thEdition$" are vertically aligned in figure 4.20. This is because, although the relative positions of the nodes are not changed, the finetuning phase of the layout, which determines the final x- and y-coordinates by aligning a node with its predecessors/successors, is still carried out.

A third example, showing the call graph of a program which converts files from one text formatting system to another is shown in figure 4.21 (25 nodes, 50 edges). Figure 4.21 shows the graph before the change. Consider the addition of an edge from "$main$" to "$fprintf$". Figure 4.22 shows the resulting graph when no layout stability is used. Figure 4.23 shows the resulting graph when layout stability is used.

Note that the graph looks substantially different between the "before" version and the "instable after" version. In contrast, the difference between the "before" and the "stable after" versions is small. In an interactive editing session, surely the "stable" version would be preferable so that the users could maintain their orientation in the graph. In addition, the number of crossings and the layout times often make the stable version preferable.

The positive results demonstrated by these figures are supported by the set of measurements shown in tables 4.2 through 4.4. These tables compare the layout times and the edge crossing counts for the three graph examples with and without layout stability. Of the four possible structural changes to a graph (add or delete a node or an edge), edge additions are chosen to measure stability. Deletion is not chosen because it could be accomplished by "hiding" those nodes or edges and not doing a new layout at all. Node addition is not chosen because the addition of a new node alone will not affect the layout of the graph until that node is connected to the graph. For each of the preceding three graphs ("World", "UNIX", and "W2t"), a pseudo-random number generator was used to select ten pairs of nodes in the graph and then an edge was added between the source and target node of each pair. The tables show the layout time and the count of edge crossings before the change, after adding the edge without layout stability, and after adding the edge with layout stability. The next to last row shows the average values for the ten randomly selected edges. The final row shows the average values when all edges are considered[4]. The measurements for the "World" graph were done on a Sun 4 workstation[5].

---

[3]The layout of this graph, originally provided by [DC], by the DAG system was shown in figure 3.9.

[4]The number of edges for each example is given by $(\sum_{i=1}^{n-1} i) - e$, where n is the number of nodes and e the number of edges.

[5]A Sun 4 workstation was not used for all three examples because the layout times were often less than one second, so the accuracy could not be measured so exactly. Furthermore, the point is to compare the layouts within each example graph, not to compare across example graphs.

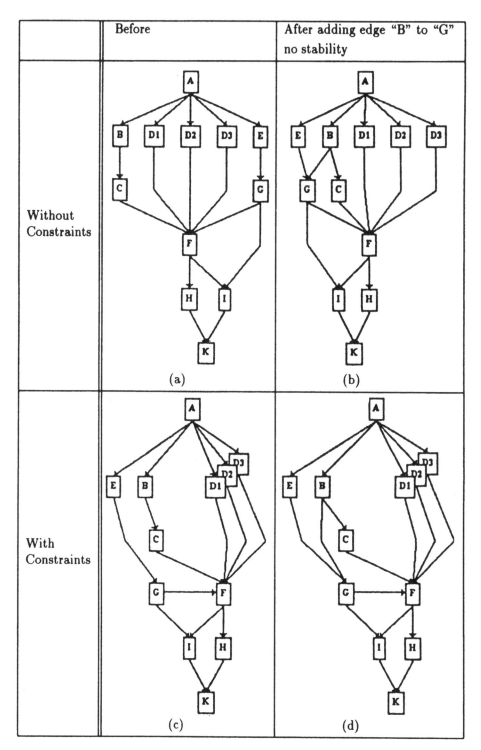

Figure 4.12: Examples of layout constraints and instability

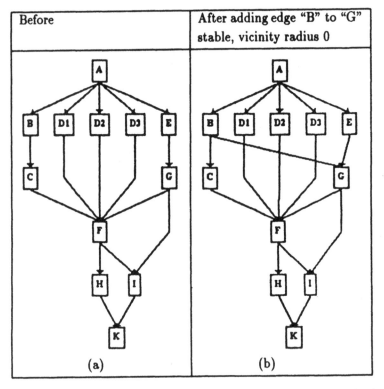

Figure 4.13: Example from 4.12, but with stability (radius 0)

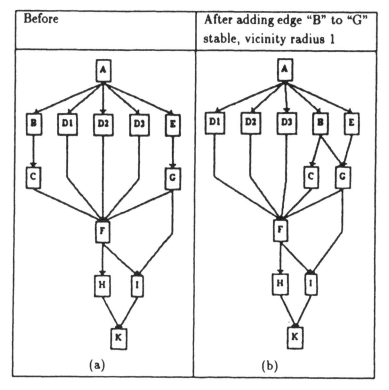

Figure 4.14: Example from 4.12, but with stability (radius 1)

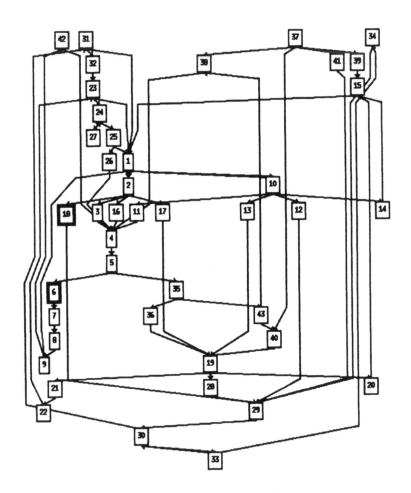

Before: 6 seconds, 40 crossings

Figure 4.15: "World" example of layout stability: Before

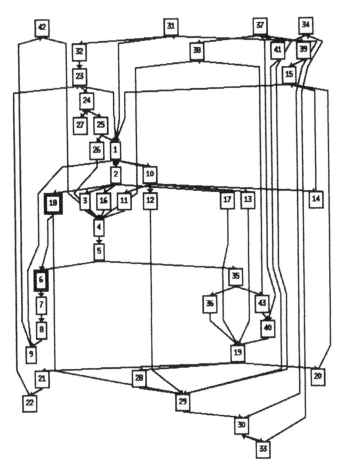

Instable layout
After adding edge from "*18*" to "*6*"
5 seconds, 45 crossings

Figure 4.16: "World" example of layout stability: Instable

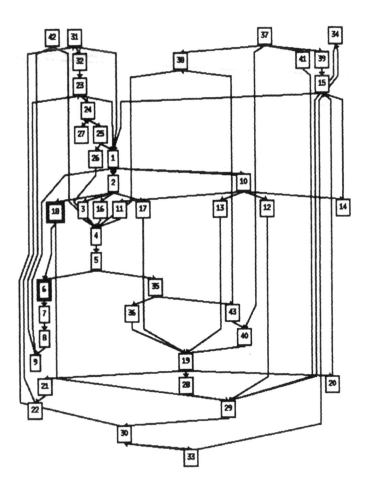

Stable layout (vicinity radius 0)
After adding edge from "*18*" to "*6*"
2 seconds, 40 crossings

Figure 4.17: "World" example of layout stability: Stable

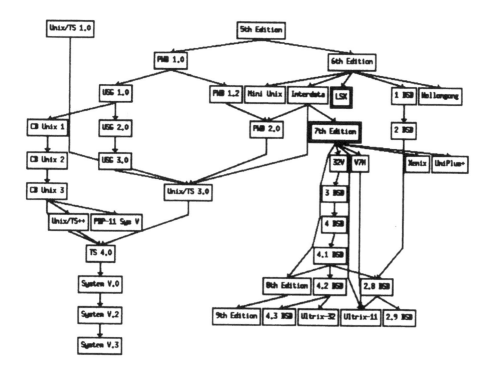

Before: 5 seconds, 5 crossings

Figure 4.18: "UNIX" example of layout stability: Before

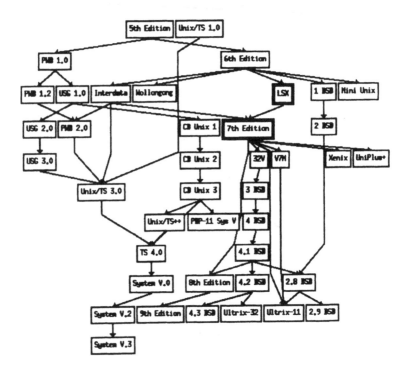

Instable layout
After adding edge from "*LSX*" to "*7th Edition*"
5 seconds, 12 crossings

Figure 4.19: "UNIX" example of layout stability: Instable

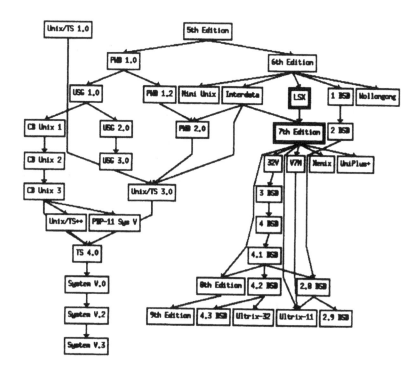

Stable layout (vicinity radius 0)
After adding edge from "LSX" to "$7thEdition$"
3 seconds, 5 crossings

Figure 4.20: "UNIX" example of layout stability: Stable

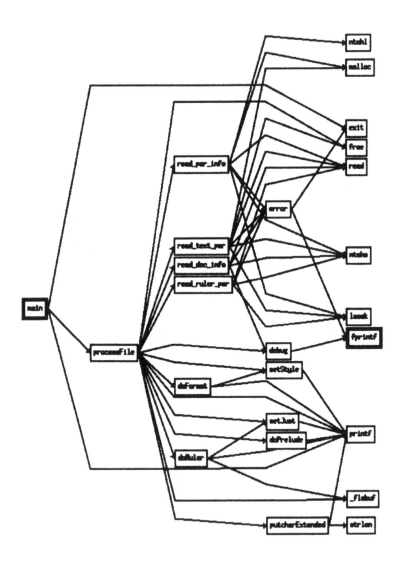

Before: 5 seconds, 70 crossings

Figure 4.21: "W2t" example of layout stability: Before

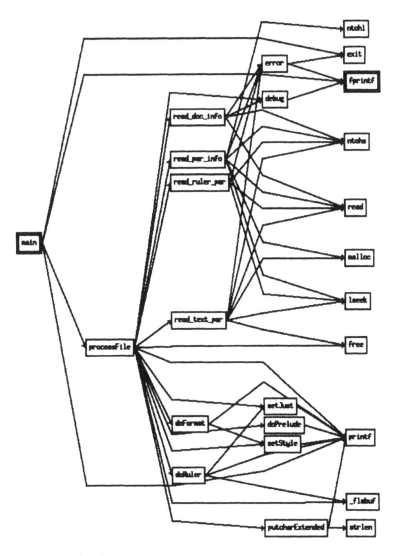

Instable layout
After adding edge from "*main*" to "*fprintf*"
5 seconds, 63 crossings

Figure 4.22: "W2t" example of layout stability: Instable

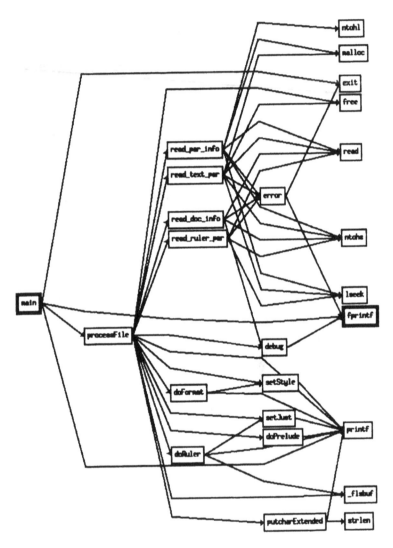

Stable layout (vicinity radius 0)
After adding edge from "*main*" to "*fprintf*"
5 seconds, 76 crossings

Figure 4.23: "W2t" example of layout stability: Stable

4.3.2 Results

There are two aspects to judging the success of layout stability: how well it performs and how efficient it is. The above examples demonstrate that layout stability works well for small and medium sized graphs. For larger graphs it is expected that the effect of layout stability will be even more dramatic. Unfortunately, it is rather difficult to measure how well the layout stability works because such a measurement is subjective. No detailed measurements, for example measuring the distances that each node moves after a change, have yet been made.

The measurements shown above provided results which were more positive than expected. They show that, not only can the user have a stable layout, but the layout can sometimes be faster and have fewer crossings than an instable layout.

The speed of the stable layout is strongly affected by the level assignment of the nodes. The time-consuming portion of the layout (the barycentric ordering phase) only has to be performed on levels which may have changed (i.e. the levels which the two nodes are on and the levels between them). This implies that if there is only a small difference in the level numbers of the nodes, then large portions of the graph will remain stable. Although the system performing a stable layout has to generate additional constraints, determine the vicinity of change, deactivate constraints in the vicinity of change, and perform a new (stable) layout, the time savings of not having to perform the barycentric ordering on a number of levels of the graph often compensates for this. It should also be noted that, although the generated set of stability constraints may be quite large, they do not introduce any additional inconsistencies.

The poor results in the "W2t" example can be attributed to two factors. Firstly, that the graph is short and wide compared to the other two examples. This makes it less likely that a large set of levels will remain stable. The second reason is that four of the randomly selected edges are between nodes which are on the same level. Recall, from section 4.2.4, that, in the case of such "flat" edges, additional constraints are generated which state that the two nodes should be neighbors. The four "flat" edges are exactly those with the longest stable layout times. And, again because of its short, wide structure, this example will have a larger proportion of "flat" edges than the two other graphs.

The fact that the number of edge crossings is often less in the stable layout is a result of the randomness of the instable layout. Because the vicinity radius in these examples is 0, all nodes will keep their positions and the only additional edge crossings will be those introduced by the additional edge. An instable layout may (or may not) result in a layout with fewer crossings. Naturally, this also implies that the use of a stable layout algorithm will inhibit the improvement of a poor layout.

	Before		Instable Layout		Stable Layout	
Edge	Time	Cross.	Time	Cross.	Time	Cross.
18 → 6	6	40	5	45	2	40
3 → 42	5	40	6	31	9	45
21 → 40	5	40	5	52	3	44
16 → 41	5	40	6	48	8	47
29 → 1	5	40	5	48	5	47
23 → 9	5	40	6	42	6	46
41 → 9	5	40	5	49	6	49
17 → 35	6	40	5	46	3	40
13 → 12	5	40	6	75	2	40
16 → 6	6	40	6	55	4	44
Average (10)	5.3	40	5.5	49.1	4.8	44.2
Average (all)	5.3	40	6.1	50.9	4.4	44.4

Table 4.2: Stable vs. instable layout: edge additions ("World")

	Before		Instable Layout		Stable Layout	
Edge	Time	Cross.	Time	Cross.	Time	Cross.
LSX → 7th Edition	5	5	5	12	3	5
System V.2 → PWB 1.0	5	5	5	7	5	6
CB Unix 2 → 4.3 BSD	5	5	5	13	5	13
UniPlus+ → 9th Edition	4	5	6	6	5	12
PDP-11 Sys V → 5th Edition	5	5	5	9	5	8
3 BSD → PWB 1.2	4	5	6	11	9	8
8th Edition → 4.3 BSD	4	5	5	5	3	5
Xenix → 8th Edition	5	5	5	6	4	11
PWB 2.0 → 7th Edition	4	5	4	8	9	5
UniPlus+ → Wollongong	5	5	4	10	9	5
Average (10)	4.6	5	5.0	8.7	5.7	7.8
Average (all)	4.7	5	5.1	9.9	4.9	8.3

Table 4.3: Stable vs. instable layout: edge additions ("UNIX")

Edge	Before		Instable Layout		Stable Layout	
	Time	Cross.	Time	Cross.	Time	Cross.
main → fprintf	5	70	5	63	6	76
lseek → read	5	70	4	60	17	70
malloc → strlen	5	70	5	57	23	70
ntohs → read_ruler_par	6	70	5	66	5	74
free → main	5	70	5	57	6	78
ntohs → processFile	5	70	5	79	6	79
debug → setStyle	4	70	6	59	13	70
fprintf → error	6	70	5	71	3	77
setJust → doFormat	5	70	7	70	10	71
flsbuf → strlen	5	70	4	65	16	70
Average (10)	5.1	70	5.1	64.7	10.5	73.5
Average (all)	5.3	70	6.4	69.8	9.8	75.6

Table 4.4: Stable vs. instable layout: edge additions ("W2t")

is a result of the randomness of the instable layout. Because the vicinity radius in these examples is 0, all nodes will keep their positions and the only additional edge crossings will be those introduced by the additional edge. An instable layout may (or may not) result in a layout with fewer crossings. Naturally, this also implies that the use of a stable layout algorithm will inhibit the improvement of a poor layout.

Chapter 5

> *"The intelligent use of equivalent forms is the touchstone of logical insight."*
> – S. K. Langer

Graphical Abstraction

Graphical abstractions provide alternative or additional representations of the information which help the user understand the graph. They are called "abstractions" because they abstract away detail which is not currently relevant to the user. Graphical abstractions provide an "in between" solution to viewing the graph at the node/edge level, where the user can only view a portion of the graph, and the layout overview, where the user can see the entire graph, but not details of the individual nodes.

The topic of graphical abstraction can be subdivided into two parts – the representation of abstractions and their definition. *Representation* refers to how the graphical abstractions are presented to the user. *Definition* refers to how the user specifies the (sub)graph to be used for the graphical abstraction.

This chapter describes three representations for graphical abstraction: hierarchical subgraph abstractions, separate views, and edge concentrations. Particular emphasis is placed on how the graphical abstractions can be defined automatically.

5.1 Related Work

Existing facilities for representation include:

- **View:** A *view* is the display of a subgraph of the graph, usually in a separate window on the screen. Several systems (e.g. GMB [JG89], GraphEd [Him88], and GraphView [BS89]) support multiple views and allow the user to cut and paste subgraphs between these views.

- **Hierarchical structuring:** Some systems (e.g. GINCOD [BNTT85], GMB [JG89]) allow the user to "zoom in" or "zoom out" of levels of abstraction.

- **Restriction:** This approach restricts the initial display to a portion of the graph and incrementally displays more (or less) as the user requests. This approach is used, for example in Prospec [CL88] and Kb-edit [TW87].

- **Navigation aids:** Some graphs are so complex that the user can get "lost" in the graph. Navigation aids help users maintain their orientation. One approach, called *focusing*, allows the user to specify a particular graph object (either directly by selecting it, or indirectly by specifying its name) and the graph is automatically scrolled so that the selected graph object appears in the middle of the scrollable region. Another popular approach is to provide a *layout overview* of the the entire graph superimposed by a placemarker box indicating the current position of the scrollable region. The use of a layout overview helps users maintain their orientation in the graph. Furthermore, in some systems the user can scroll the graph by moving the placemarker box. Such a continuous scrolling mechanism is far preferable to reliance on independent horizontal and vertical scrolling mechanisms.

In almost all cases, certainly in all of the systems described in chapter 3, the subgraph abstraction is defined manually by the user. Facilities to automatically define the subgraph used in the graphical abstraction are of particular interest here and this makes work in the area of graph clustering relevant. The ARCH project [SP89] uses a clustering method based on cross-references to automatically analyze the structure of existing systems. The goal is to rediscover the structure of a system that has been modified over the years and compare it with the intended structure. The clustering tool presents the user with a summarized view of the call graph. In [MG89] a graph layout is performed by partitioning the graph into relatively independent subgraphs which are then laid out separately. It is suggested there that the user could choose to view the subgraph in its entirety or as a "summary" node representing the subgraph. The display of hierarchical subgraph abstractions within a directed graph is particularly complex. In [SM91] Sugiyama and Misue present an algorithm for drawing a compound digraph, a graph containing both inclusion (hierarchy) and adjacency edges. An example is shown in figure 5.1. The user can reorganize the structure by expanding, abbreviating, releasing, or grouping nodes.

In [Sny80] a method for storing sparse, directed graphs with unlabeled edges in a particularly compact format is presented. The proposed method partitions the edges of the graph into complete bipartite subgraphs and stores this more compact description. By using this technique, a fifty to seventy percent reduction in storage space is achieved. Although the area of application, lookahead sets of large LR parser decision functions, is not particularly relevant to graph

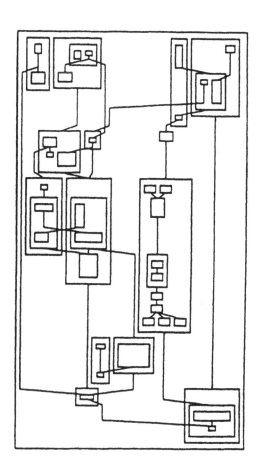

Figure 5.1: Display of a compound digraph (from [SM91])

editors, this compact machine representation has many similarities with the edge concentration method presented in section 5.3.

5.2   Subgraph Abstraction

A *subgraph abstraction* is a subgraph with one or more graphical representations. The user can choose which graphical representation should be used to display the subgraph and can thus control the level of detail shown. When viewing a graph, subgraph abstractions can be used to hide subgraphs that are not currently of interest to the user. Alternatively, a subgraph abstraction can be used to give the user an uncluttered display of the subgraph alone. In either case, the subgraph abstraction is used to focus the user's attention on the relevant portions of the graph.

5.2.1   Representation

Hierarchical Subgraph Abstractions

The subgraph abstraction proposed here represents a subgraph as a node in the graph. This subgraph abstraction node appears in the context of the rest of the graph and is positioned by the layout algorithm just like other nodes in the graph.

The subgraph abstraction can be viewed in several different ways. It can be displayed as a normal-sized node with a special icon denoting that it represents an abstraction. This is called a "black-box" view of the abstraction because the user cannot see the subgraph which is "inside" the abstraction node. Alternatively, the abstraction node can be displayed as a large node and the subgraph is shown within the confines of the node. In either case the abstraction node is shown within the context of the rest of the graph and this raises the question of how to represent edges which cross abstraction boundaries. Two alternative representations are used to display such edges – in the first case edges from the outside stop at the abstraction's boundary and in the second case edges are drawn all the way from source node to target node. These two representations are called "grey-box" and "white-box" view respectively (the user can see partially or all the way "into" the abstraction node.). The user can switch between these different representations easily. Figure 5.2 shows a graph representing a one-bit adder. The subgraph representing the "exclusive or" operation is a hierarchical subgraph abstraction shown as a "black-box", "white-box" and "grey-box" view. As can be seen in figures 5.2 and 5.4, showing a subgraph abstraction in the context of the rest of the graph can lead to sub-optimal use of the screen "real estate". The Sugiyama layout (see section 4.1.4) used for these examples spaces the levels based on the height of the tallest node in the level.

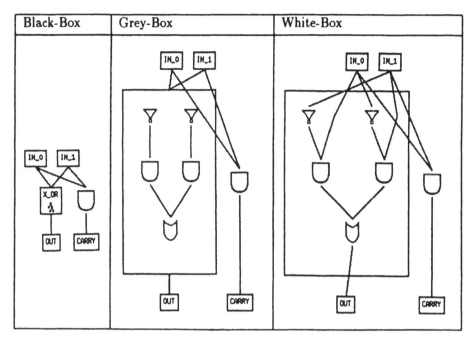

Figure 5.2: Black-, grey-, and white-box views of a subgraph abstraction

This may force nodes on subsequent levels to be placed further away from their predecessors than necessary. A layout algorithm which better takes different sized nodes into account could, however, avoid this drawback.

One of the design decisions for the subgraph abstractions is that the layout of the subgraph is entirely independent of the layout of the rest of the graph. This speeds up the layout of the graph because only (sub)graphs affected by a change have to be laid out. This does, however, lead to a minor problem in the "white-box" view. Because the layout inside and outside of a subgraph abstraction is done independently, the source and target nodes of an edge which crosses an abstraction boundary are not part of the same layout. In a "white-box" view, such edges are drawn in after the layout has been performed. This can lead to unnecessary crossings as is shown in the following example.

Consider the graph shown in figure 5.3(a). After adding an edge from Z to A, a new layout of the outer graph is performed. Conceivably (but not necessarily) the resulting layout would change the relative positions of the nodes X and Y, and thus the relative positions of nodes A and B. If that happens, then there will be an additional, unnecessary crossing in the graph as shown in figure 5.3(b). The user can avoid the issue by using the "grey-box" view or the user could influence the layout by using layout constraints to fix the layout (as shown in figure 5.3(c)).

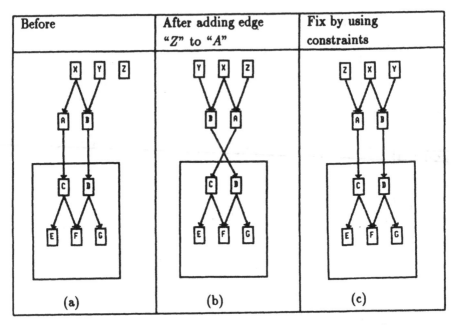

Before	After adding edge "Z" to "A"	Fix by using constraints
(a)	(b)	(c)

Figure 5.3: Adding edge causes unnecessary edge crossing

The design choice of representing the subgraph as a node in the context of the graph allows the user to build up a hierarchy of subgraph abstractions. The hierarchy is *strict*, meaning that one node may only be (directly) in one subgraph abstraction. Edges between nodes in different abstractions belong to the innermost common (sub)graph. Figure 5.4(a) shows a graph representing the series of assembly/disassembly operations a robot undertakes to put together a simple object. Figure 5.4(b) shows the same graph using four levels of abstraction. At the innermost level, the two nodes for part E are grouped into a "black-box" abstraction. This is part of a "grey-box" abstraction which is part of a "white-box" abstraction. At the outermost level, the entire graph is also represented as a subgraph abstraction. The user can change the viewing status of any of these abstractions to hide details or to show more detail.

Separate View

A *separate view* is a graph editing session on a subgraph of the graph being edited. It is shown in a separate window on the screen rather than in the context of the graph. The use of multiple, separate views is a powerful technique because it allows the user to keep several graph views "active" simultaneously. They allow the user the freedom to experiment and easily return to a known, stable state. A separate view can also, for example, be used to display a graph containing different edge types as a set of separate views, one for each edge type.

Changes made to the graph in the separate view are not automatically re-

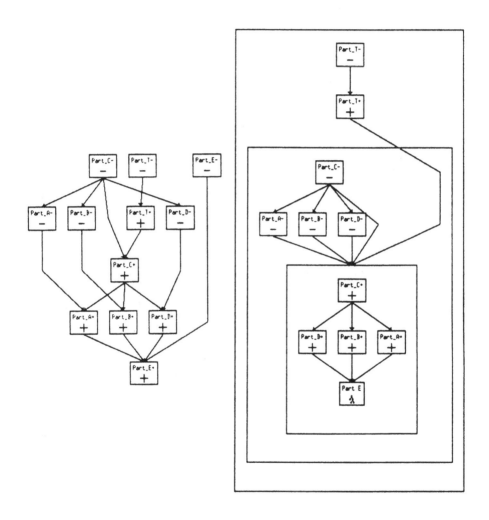

(a) Before   (b) After creating hierarchical abstractions

Figure 5.4: Multi-level hierarchical subgraph abstraction

flected in the original graph. Conceivably, there could be several separate views containing possibly different representations of the same node and only the user can decide which changes should be made to the graph in the main editing session. However, facilities which make the update operation easier for the user are available. For example, the user can choose between taking over the changes in the separate view or merging the changes with the original graph. Further information on the separate view will be presented in section 8.2.1.

5.2.2 Definition

This section describes the various ways in which the subgraph can be specified. Unless otherwise noted, the subgraph will be defined by specifying the set of nodes in the subgraph. The edge set of the subgraph is defined to be any edges which have both source and target nodes in the subgraph.

- **Manual (individual selection):** The user selects the subgraph manually by selecting a set of nodes with the pointing device (e.g. by dragging a "rubberband"). A hierarchical subgraph abstraction can be selected just like any other node.

- **Manual (area selection):** The user selects a subgraph by specifying an area on the screen with the pointing device. The subgraph could be refined by adding and deleting individual nodes and edges from this selection.

- **Typename:** The user specifies a set of typenames, and nodes and edges of this type define the subgraph.

- **Regular expression:** The user specifies a regular expression, and nodes whose title matches the regular expression define the subgraph.

- **Transitive closure:** The user specifies a node, and all successors (or predecessors) of this node define the subgraph. The user could additionally specify that only edges of a particular type should be followed.

- **Read from file:** The user can read a description of the graph (e.g. in GRL) from a file. The origin of this file could be, for example: 1) a graph description saved from a prior editing session, 2) a graph description created by the user using a text editor, or 3) a graph description generated by some other program.

- **Function:** The user specifies the name of an application-specific function which, given the graph as an argument, returns a subgraph. For example in a configuration management application, the application developer could provide a function which identifies all bug reports for a particular version of the software that are over four weeks old.

Certainly there are further possibilities for defining subgraph abstractions, but this set meets most users' basic needs and is extendible via the "function" alternative.

5.2.3 Examples

In addition to the examples presented thus far, examples of hierarchical subgraph abstractions are shown in section 6.4 and in figure 8.7. Figure 5.5 shows an example of the separate view facility. The user has chosen to view the hierarchical subgraph abstraction "Research" in a separate view. The "Research" subgraph abstraction consists of the node "Extendibility" and a subgraph with four nodes representing research areas for graph editors. The fact that each subgraph is laid out independently allows the user to select different orientations or different layout algorithms for each subgraph. The subgraph in the separate view is shown using a left-to-right rather than a top-to-bottom orientation. A further example, where a separate view is used to show a different but related graph, is given in figure 7.2.

5.2.4 Results

Any graph editor, and especially an extendible one which will be used for many different applications, will benefit from a variety of graphical abstractions. They should all interact with the graph editor through the same interface, and be as similar as possible rather than being a set of independent, possibly incompatible, techniques. This section showed two different subgraph abstraction representations: hierarchical subgraph abstractions shown in the context of the graph and a subgraph abstraction shown as a separate view. The two approaches are compatible. For example, the user can select a multi-level subgraph abstraction, choose to view it as a separate view, and the lower-level abstractions will also be copied to the separate view. Both use exactly the same facilities for defining the subgraph.

5.3 Edge Concentration

Graphs depicting a system's configuration, call graphs, graphs depicting import and export relationships between modules, and graphs depicting the "includes" relation among a system's source files are examples of graphs which have a large number of edges in proportion to nodes. Their representation may contain so many edges that the graph is essentially useless as a visual aid to understanding the relationship between the nodes. Such graphs, due to their nature, often contain many complete bipartite subgraphs (CBSs). A technique is presented in this section which identifies such subgraphs and replaces them with an equivalent

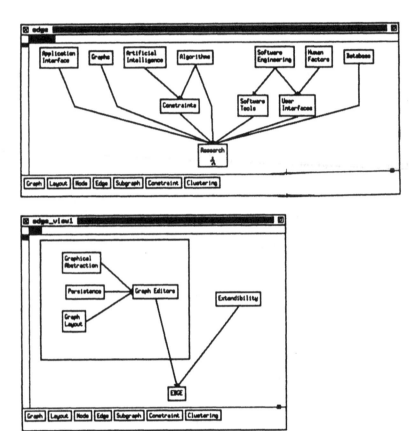

Figure 5.5: Separate view of a subgraph abstraction

tripartite representation which uses fewer edges. The layout of the resulting graph has fewer edges and often far fewer edge crossings[New89].

Note: It may be helpful to the reader to review the definitions of coverings, partitions, and bipartite, tripartite, and complete (sub)graphs presented in chapter 2.

5.3.1   Representation

Given a complete bipartite graph $B = (N1, N2, E)$, the equivalent tripartite representation $T = (N1, EC, N2, E^*)$ is constructed by inserting an additional level containing a single node (called the *edge concentration node*) between the two levels. The set of edges E^* is defined by $E^* = (N1 \times EC) \cup (EC \times N2)$. In other words, the set of edges in the complete bipartite subgraph are represented by a special node whose "fan-in" is all of the source nodes and whose "fan-out" is all of the target nodes. The set of edges of the complete bipartite subgraph

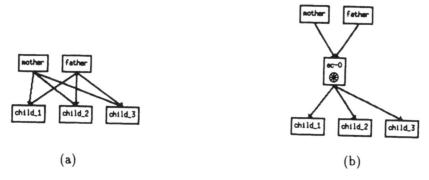

(a)   (b)

Figure 5.6: A graph and its representation using an edge concentration

(i.e. E) is called an *edge concentration* and the resulting graph is said to be *concentrated* .

The representation is equivalent because B can be obtained from T by calculating the transitive closure of T and then deleting the edge concentration node. If n is the number of nodes in $N1$ and m is the number of nodes in $N2$, then the complete bipartite graph B has $n + m$ nodes and $n \times m$ edges. In comparison, the tripartite representation T has $n + m + 1$ nodes and $n + m$ edges.

Consider the graph shown in figure 5.6(a). The set of edges { ($mother,child\_1$) ($mother,child\_2$) ($mother,child\_3$) ($father,child\_1$) ($father,child\_2$) ($father,child\_3$) } is an edge concentration. Figure 5.6(b) depicts the same graph by using the corresponding edge concentration node (labeled *ec-0* and marked with a special icon). Note that the concentrated graph has fewer edges and crossings and is more readable. In many cases no information is lost, but this depends on the relationship represented by the edges. If the relationship represented by the edge is "is-parent-of" then no information is lost. However, if the edges were of different types, i.e. explicitly representing the "is-mother-of" and "is-father-of", then this information would no longer be present. Similarly, labels or other information associated with the edges will be lost by concentrating the graph. One disadvantage of using edge concentrations is that the user now has to follow a level of indirection to locate the two endpoints of a concentrated edge, but at least the edge is not hidden in a confusing mass of crossings. The larger, more complex examples given later in the chapter will more clearly demonstrate the usefulness of edge concentrations.

5.3.2   Definition

Once the edge concentrations are identified, it is a simple task to convert them to an equivalent representation. The difficult part is to find an appropriate set of edge concentrations in an arbitrary graph. To simplify the problem, the search is restricted to a bipartite subgraph $B = (N1, N2, E')$ of $G = (N, E)$ such that $N1 \subset N$ is the set of all nodes on a particular level of G, and $N2 \subset N$ is the set of all successors of nodes in $N1$. $E' \subset E$ is the set of edges between $N1$ and $N2$. As will be shown later, this approach can be applied repeatedly to concentrate a non-bipartite graph.

The *edge concentration problem* is formally stated as follows.

INSTANCE: Given a bipartite graph $G = (N1, N2, E)$ and a positive integer K.

QUESTION: Is there a set of complete bipartite subgraphs $G_1, G_2, ... G_s$ of G $(G_i = (N1_i, N2_i, E_i), N1_i \subset N1, N2_i \subset N2, E_i \subset E)$ of G that cover all edges in G such that the sum of all edges in the equivalent, tripartite representations G_i^* of G_i (i.e. $\sum_{i=1}^{s} |E_i^*|$) is $\leq K$?

In other words, the goal is to find the set of (possibly overlapping) complete bipartite subgraphs that cover all edges of the graph such that the total number of edges in the concentrated graph is minimized.

5.3.3 The Complexity of the Edge Concentration Problem

There appears to be no polynomial time solution to the edge concentration problem, but this has not yet been proved. A similar, apparently simpler, problem known as the "covering by complete bipartite subgraphs" problem is known to be NP-complete [GJ79]. The "covering by complete bipartite subgraphs" problem can be formally stated as follows:

INSTANCE: Given a bipartite graph $G = (N, E)$ and a positive integer K.

QUESTION: Is there a set of $k <= K$ complete bipartite subgraphs $G_1, G_2, ... G_k$ $(G_i = (N1_i, N2_i, E_i), N1_i \subset N1, N2_i \subset N2, E_i \subset E)$ of G that cover all edges in G?

In other words, the goal is to find the smallest set of (possibly overlapping) complete bipartite subgraphs that cover all the edges in the graph. In contrast, in the edge concentration problem, the **number** of complete bipartite subgraphs is irrelevant. Instead, the goal is to find the set which minimizes the total number of edges in the concentrated graph. The following compares the two solutions in a concrete example.

Consider the graphs shown in figure 5.7. Figure 5.7(b) shows the graph from figure 5.7(a) with a set of complete bipartite subgraphs given by the edge concentration problem. Figure 5.7(c) shows the corresponding solution to the "covering by complete bipartite subgraphs" problem. There, the two complete bipartite subgraphs depicted by the edge concentration nodes *ec-0* and *ec-1* is

(a) Before: 23 edges, 52 crossings

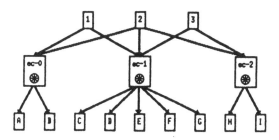

(b)Edge concentration: 16 edges, 2 crossings

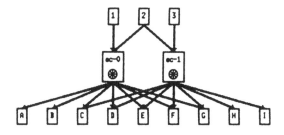

(c) Covering by CBSs: 18 edges, 10 crossings

Figure 5.7: Comparison of two coverings by complete bipartite subgraphs

the smallest number that will cover the graph. The "covering by complete
bipartite subgraphs" problem, however, does not necessarily result in a graph
with the fewest number of total edges. For example, the optimal solution shown
in figure 5.7(b) has (2+2)+(3+5)+(2+2) = 16 edges whereas the "covering by
complete bipartite subgraphs" solution shown in figure 5.7(c) has (2+7)+(2+7)
= 18 edges.

Although no reduction has yet been found from a known NP-complete problem to the edge concentration problem, the problem appears to be at least as
hard as the NP-complete problem "covering by complete bipartite subgraphs".
The difficulty in finding a transformation has to do with the fact that, in the
edge concentration problem, the subgraphs cannot be treated in isolation.

5.3.4   An Approximate Solution

This section describes the "edge concentration algorithm", a polynomial-time
approximate solution to the edge concentration problem. The algorithm identifies potential edge concentrations by considering the complete bipartite subgraph formed by each pair of source nodes and their successors. The subgraph
so formed is called an *intersection*. If the target nodes of an intersection are a
subset of the target nodes of a previously determined intersection, then the set of
target nodes is partitioned. The partitioning of the target nodes avoids the overlap in the "covering by complete bipartite subgraphs" solution (see figure 5.7(c))
which led to additional edges and crossings.

A minimum size, M, for the concentrations is specified by the user. The size
is calculated as (number-of-target-nodes – 1) * (number-of-source-nodes). This
formulation was chosen so that concentrations that have only a single target
node will never be created.

Edge Concentration Algorithm

The approximate solution to concentrate the edges from nodes on level L is as
follows:

1 Set initial list of concentrations, $CLIST$, to the empty list.
   Set initial list of intersections, $ILIST$, to the empty list.

2 Sort the successors of each node if not already sorted[1].

3 For each pair of nodes (N_1, N_2) in level L

   3.1 Construct an intersection I from N_1, N_2 and the intersection of their
        successor nodes.

---

[1]Successors are sorted once to make the calculation of the intersections easier in the next
step.

3.2 Add I to a list of intersections called $ILIST$.

4 For each intersection I in $ILIST$

    4.1 If the size(I) < M then discard I and continue loop 4.

    4.2 Compare the intersection I with each concentration C in $CLIST$
        * if target-nodes(I) = target-nodes(C) then
           · Add nodes in source-nodes(I) to the nodes in source-nodes(C).
        * Continue loop 4.

    4.3 Compare the intersection I with each concentration C in $CLIST$
        * if target-nodes(I) $\supset$ target-nodes(C) then
           · Split I into I' and C'. C' contains the target-nodes(I) that are also in target-nodes(C). I' contains the remaining target-nodes(I).
           · Add I' and C' to $CLIST$.
           · Continue loop 4.

    4.4 Compare the I with each concentration C in $CLIST$
        * if target-nodes(I) $\subset$ target-nodes(C) then
           · Split C into I' and C'. I' contains the target-nodes(I) that are also in target-nodes(C). C' contains the remaining target-nodes(I).
           · Replace C by I' and C' in $CLIST$.
           · Continue loop 4.

    4.5 Add I to $CLIST$.

5 Merge concentrations that have the same set of target nodes and discard those whose size is < M.

A Simple Example

For example, consider the graph in figure 5.7(a). The target nodes of the pairs of intersections are: {{CDEFG} {ABCDEFG} {CDEFGHI}}. Adding these one-by-one to an initially-empty list of concentrations:

- Add concentration {CDEFG}. Concentration list is now {{CDEFG}}.

- Add concentration {ABCDEFG}. Find that existing concentration {CDEFG} is a subset, so split concentration {ABCDEFG} into {AB} and {CDEFG} and add each. Concentration list is now {{CDEFG} {AB}}.

- Add concentration {CDEFGHI}. Find that existing concentration {CDEFG} is a subset, so split concentration {CDEFGHI} into {CDEFG} and {HI} and add each. Concentration list is now {{CDEFG} {AB} {CDEFG} {HI}}.

• After merging, concentration list becomes {{CDEFG} {AB} {HI}}.

Thus the list of edge concentrations that is the same as the optimal set shown in figure 5.7(b) is obtained.

A Counter Example

The suggested heuristic of splitting the intersections when one is found to be a subset of another may not always be optimal, as is shown in the following example. Figure 5.8(a) shows the original bipartite graph, Figure 5.8(b) shows the optimally concentrated graph, and figure 5.8(c) shows the concentrated graph generated by the approximation algorithm. The problem arises because many nodes in the top level go to the set {ABCDE}, but that concentration is split because two nodes also go to the subset {AB}.

Runtime of the Edge Concentration Algorithm

Given a bipartite graph $G = (N1, N2, E)$ where $n = |N1|$ is the number nodes in the top level and $m = |N2|$ is the number of nodes in the bottom level and given that the value k is the number of concentrations found. Sorting the successors of each node requires $O(m\log m)$ and since there are n nodes in the top level, this gives us $O(nm\log m)$ for Step 2. Calculating the intersection takes $O(m)$ (since successors are already sorted). The intersection will be calculated for each pair of nodes, thus $\frac{mn(n-1)}{2}$ operations are required to generate the list of potential concentrations in Step 3. Step 4 will be executed once for each of the $\frac{n(n-1)}{2}$ concentrations. Initially, the list of concentrations to compare each concentration against will be empty, but each time through the list the list of concentrations may grow by at most two. Therefore, the number of comparisons in Step 4 is bounded by $\sum_{i=0}^{\frac{n(n-1)}{2}} 2i$ or $O(n^4)$. The final result is $O(nm\log m + \frac{mn(n-1)}{2} + n^4)$, which, assuming that $m < n^2$, gives us $O(n^4)$.

It should be noted that the worst case is much worse than cases likely to be encountered in practice. The above worst-case scenario assumes that some distinct non-empty intersection will be generated for every pair of nodes in the top level. In practice, a large number of the intersections are empty or the same as the intersection generated for a different pair of nodes. Thus, the set of concentrations generated in Step 4 is typically much smaller than the worst case would allow.

Concentrating Arbitrary Directed Graphs

The edge concentration algorithm can also be applied to non-bipartite directed graphs by applying the algorithm repeatedly to various levels in the graph. The

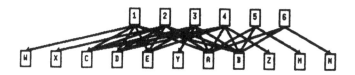

(a) Before: 30 edges, 87 crossings

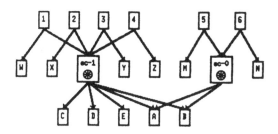

(b) Optimal solution: 19 edges, 3 crossings

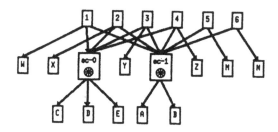

(c) Approximate solution: 21 edges, 12 crossings

Figure 5.8: Counter example

user must specify not only the size of the concentrations, but also the choice
of which level to concentrate. Typically the user chooses the level with the
most crossings as the one to concentrate. The list of potential concentrations is
calculated and displayed to the user upon request. The user then decides whether
to use that set of concentrations or to try a different level or concentration size.
After the user has concentrated a level, the modified graph is laid out again.

Figure 5.9 shows an example of a graph which initially has 20 edges and 36
crossings. After concentrating level 0 with a minimum size of 4 the number of
edges was reduced to 15 and the number of crossings was reduced to 6. After
further concentrating level 0, this time with a minimum size of 2, the number of
edges was reduced to 14 and all crossings were eliminated.

5.3.5 Examples

The following set of "real-life" examples demonstrate the effectiveness of using
edge concentrations. In each case applying the edge concentration algorithm two
times was sufficient to cause a significant reduction in the number of crossings.
The Sugiyama layout algorithm (see section 4.1.4) is used to lay out all of the
following graphs[2].

---

[2]Although the Sugiyama layout algorithm attempts to reduce the number of crossings in
the graph, it does not necessarily find the minimal number of crossings. Therefore, it may be
possible to display these graphs with even fewer crossings.

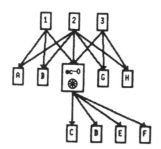

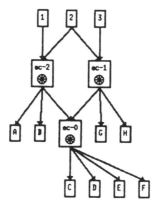

Figure 5.9: Series of edge concentrations

Example of includes relation (Texchk program)

Texchk is a program that tests a TeX/LaTeX file for correct syntax. The "Before" graph at the top of figure 5.10 shows the includes relationship generated by the makedepend program[3]. The sequence of edge concentration commands used to arrive at the lower picture was:

      Concentrate level 0, minimum size 6.

      Concentrate (the new) level 1, minimum size 4.

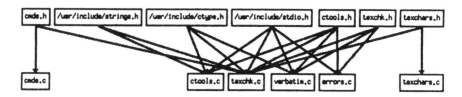

Before: 20 edges, 33 crossings

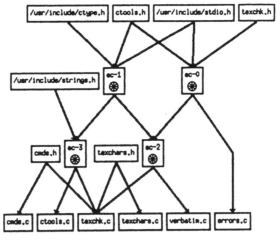

After: 19 edges, 3 crossings

Figure 5.10: Texchk program – includes relation

---

[3]Makedepend is a program that generates a list of dependencies between source and header files.

Example showing definition and use of variables (Calls program)

Calls is a program that takes source code in the C language and produces the corresponding call graph. The "Before" graph at the top of figure 5.11 shows the definition and use of variables used in the program. The sequence of edge concentration commands used to arrive at the lower picture was:

Concentrate level 0, minimum size 14.
Concentrate level 0, minimum size 10.

Example of calls relation (W2t program)

W2t is a program that translates MacWrite files to Troff. The "Before" graph in figure 5.12 shows the calls relationship among the procedures. The sequence of edge concentration commands used to arrive at the "after" picture shown in figure 5.13 was:

Concentrate level 2, minimum size 10.
Concentrate (the new) level 3, minimum size 8.

Example of includes relation (Xcal program)

Xcal is an X Windows System calendar program. The "Before" graph at the top of figure 5.14 shows the includes relationship between the source and header files for the program. The sequence of edge concentration commands used to arrive at the lower picture was:

Concentrate level 0, minimum size 4.
Concentrate level 0, minimum size 4.

Example of includes relation (Fig program)

Fig is a drawing program for use under SunWindows. The "Before" graph at the top of figure 5.15 shows the dependencies which were specified by the author in the Makefile. The sequence of edge concentration commands used to arrive at the lower picture was:

Concentrate level 0, minimum size 14.
Concentrate (the new) level 1, minimum size 8.

Before: 62 edges, 230 crossings

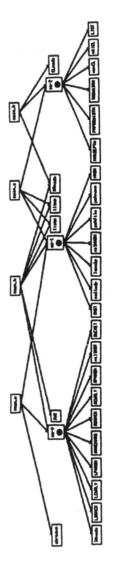

After: 39 edges, 9 crossings

Figure 5.11: Calls program – define/use relation

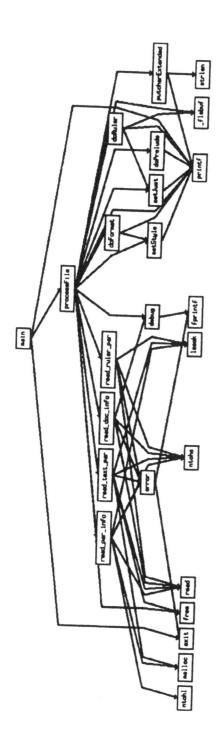

Before: 50 edges, 70 crossings

Figure 5.12: W2t program – calls relation (before)

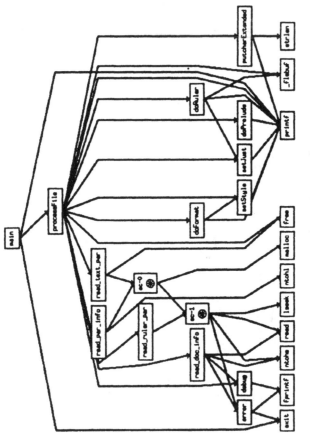

After: 45 edges, 14 crossings

Figure 5.13: W2t program – calls relation (after)

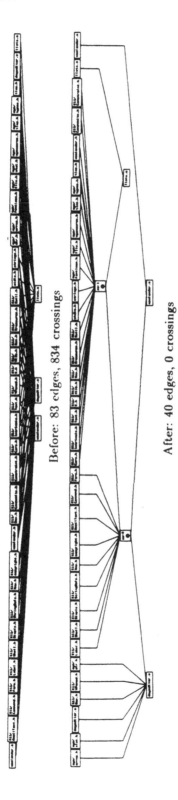

Before: 83 edges, 834 crossings

After: 40 edges, 0 crossings

Figure 5.14: Xcal program – includes relation

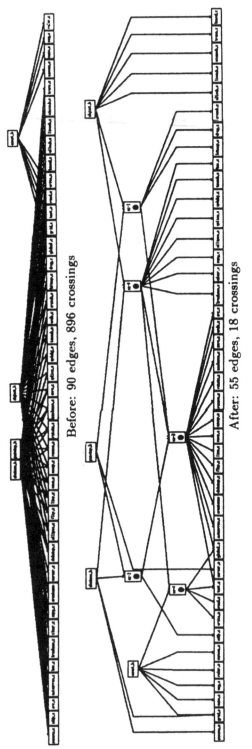

Before: 90 edges, 896 crossings

After: 55 edges, 18 crossings

Figure 5.15: Fig program – includes relation

5.3.6   Results

Table 5.1 summarizes the results from the five examples. The table shows how
the number of edges and the number of edge crossings changes after each pass
of the edge concentration algorithm. The column labeled "% Elim." shows
what percentage of the crossings were eliminated after two passes of the edge
concentration algorithm. The "Time" column shows the total amount of time
taken by both passes of the edge concentration algorithm measured on a Sun 3
workstation.

Name	Before		After 1st Pass		After 2nd Pass		% Elim.	Time
	Edges	Cross.	Edges	Cross.	Edges	Cross.		
Texchk	20	33	18	9	19	3	91%	2 sec.
Calls	62	230	43	29	39	9	96%	2 sec.
W2t	50	70	47	40	45	14	80%	2 sec.
Xcal	83	834	57	171	40	0	100%	4 sec.
Fig	90	896	82	436	55	18	98%	4 sec.

Table 5.1: Effectiveness of edge concentration algorithm

It is clear from the above table that the use of the edge concentration algorithm can cause significant reduction in the number of crossings in a graph. The
number of edges is not significantly affected, but the addition of edge concentrations does tend to spread the graph out vertically and make it easier to read. A
further advantage of this clustering technique is that it is helpful as a "reverse
engineering" tool – it can automatically detect nodes that are strongly related
and display them as such. This can be useful for familiarizing a new person with
the project or to compare the actual program against specification documents.

It should be noted that the edge concentration algorithm produces dramatic
results in some cases where poor programming style was used. In a further example (not shown above) there were 78 nodes, 734 edges, and 117,876 crossings in
the original graph representing the "includes" relation. The addition of a single
edge concentration node dropped the number of crossings to 61. Examination
of the source code from which the graph was obtained showed that all of the
source files included the file "all.h", and "all.h" included a dozen other header
files. This is a good example of where the edge concentrations can be useful for
"reverse engineering" by pointing out a problem area.

Chapter 6

Persistence

When using an interactive tool, a user would like to be able to save the current state of the system and be able to restart later with exactly the same configuration. In general, what the user really wants is for some, if not all, of the tool's data structures to be saved in non-volatile storage in some *external representation* that can later be reloaded. The data structures can thus be preserved beyond the execution of the program that created them, and this is the definition of *persistence* that will be used here. The two main alternatives for the storage of persistent data are files and databases. Databases are appropriate when concurrency and high reliability are important. In the case of the "session persistence" described above, a solution based on files is sufficient and this is the choice presented in this chapter. In either case, it is important that the user does not (have to) care about how the persistence is realized.

This chapter describes a language which can be used for saving an external representation of the graph and the editing session. The following is a list of desirable properties of such a language.

- **Complete:** The representation should be able to record all relevant information about the graph and the editing session. In particular, the following information should be represented.

  - The structure of the graph. This includes the list of nodes and edges and any graphical abstractions.

  - Attributes affecting the display of the graph. This includes attributes such as the choice of layout algorithm, the layout constraints, default display attributes for node or edge types, the color of a particular edge etc..

  – The status of the editing session. This includes the current window
    size and position, the scrolling position within the window, the set of
    currently selected nodes etc..

- **Extensible:** Application-specific attributes should be represented in the
  same manner as the standard set of attributes.

- **Modular:** It should be possible to split the external representation into
  independent segments. In particular, it is desirable to be able to extract
  the description of a subgraph so that it can be used independently of the
  rest of the graph.

- **Editable:** It should be possible for the user to edit the external representation.

- **Flexible:** The ordering of the entries should not be fixed and it should be
  possible to omit portions of entries.

- **Independent:** The same format should be usable across all machines,
  systems, and implementation languages.

6.1   Related Work

6.1.1   Overview of Methods to Achieve Persistence

Persistence is concerned with how long a variable exists. The following listing,
quoted from [ABC+83], shows the spectrum of persistence:

1. transient results in expression evaluation

2. local variables in procedure activations

3. own variables, global variables, and heap items whose extent is different
   from
   their scope

4. data that exists between executions of a program

5. data that exists between various versions of a program

6. data that outlives the program

Persistence in the first half of the list is typically provided by a programming
language whereas persistence in the second half of the list is typically supported
by files or a database. The solution proposed here is designed to meet the fourth
category of persistence, but some support for the problem of various versions of
the program is also provided.

Persistence is a desirable goal in a wide variety of systems and various methods have been proposed to achieve persistence (see also [BM92]). In the simplest
approaches, where the persistent storage is a backup for the data structures in
memory, files are sufficient. In the more sophisticated approaches, where the data
structures in memory are used as a cache for the persistent storage, databases
are more appropriate. A summary of various methods in increasing level of
sophistication follows (see also [CM88]).

- **Snapshot:** This "all or nothing" version of persistence saves a copy of the
  core image of the program in secondary storage. This image can later be
  reloaded. This approach is only useful for restarting the same program on
  the same type of machine, because the core image which is saved is not
  independent of these factors. A further disadvantage is that this approach
  is that it is not modular – one can only save the whole image or nothing
  at all.

- **Textual Flattening:** This approach encodes the data structures in a
  textual representation which can then be stored in secondary storage. Although this method involves a substantial overhead in converting the data
  structures to and from a textual representation, the advantages of this
  method over the snapshot approach is that it is machine, language, and
  program independent. Furthermore, since the representation is kept in
  textual form in a file on disk, it is possible for the user to edit it[1]. A
  disadvantage of this approach is that different methods must be used to
  generate the textual representation of each data type. In particular, any
  solution based on textual flattening must also address the issue of how
  pointer variables will be represented.

- **Multiple Address Space Model:** In this approach, the system maintains an object table containing an entry for each persistent data structure.
  The object table maintains a mapping between a persistent data structure's
  "PID" (for persistent identifier) and its address (either memory address or
  disk address). When a programmer accesses a persistent data structure
  which is not currently in main memory, the system responds to this "object fault" by loading the structure from secondary storage. Such a solution
  is said to offer *transparent* persistence because the programmer does not
  have to be aware of it. Many of the existing solutions to persistence are
  based on some version of this approach[WWFT88, RC89, ACCM83].

- **Database:** Instead of storing the data structure in a normal file, persistent data structures are often saved in a database. The advantage of this
  approach is that the structures can be shared by several programs or users.

---

[1]Note that for some application areas this is not necessarily a desirable aspect because this
gives the user the ability to make the structure inconsistent.

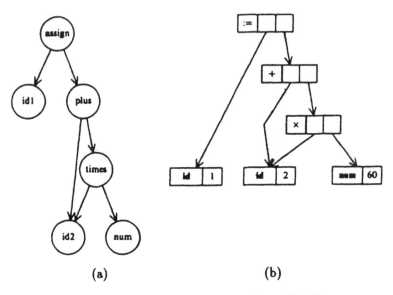

(a)   (b)

Figure 6.1: DRAG examples (from [Tri88])

6.1.2   External Formats for Graph-based Tools

- Most graph editors provide no facilities for persistence per se, but they do have an input/output format for reading and writing the graph's structure. The format is usually very simple consisting, for example, of a list of nodes and edges.

- DRAG [Tri88] allows a range of specifications. The minimum specification describes only the graph's structure, whereas the full specification also specifies how to draw each node and edge, where edges attach to nodes, etc.. Figures 6.1(a) and (b) show the same graph with the minimum and with a more detailed specification. The minimum specification is the adjacency list of the graph:

```
assign -> id1, plus;
plus -> id2, times;
times -> id2, num;
```

The detailed specification for figure 6.1(b) additionally describes the appearance of the nodes. In this example, the statement

```
node id1 pair "id" "1" rank 4;
```

is used to display node id1. This statement specifies that node id1 is of nodetype "pair" and should display the two labels "id" and "1". Addition-

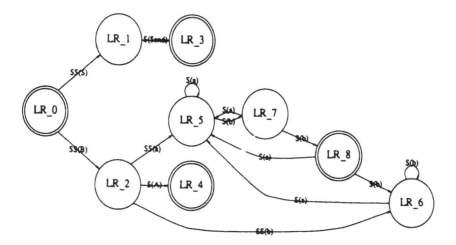

Figure 6.2: DAG example (from [GNV88])

ally the node should appear on rank 4 of the graph. "pair" is a user-defined node type whose appearance is specified by text formatting commands.

DRAG allows several other statements affecting the appearance of the graph. One can specify labels for edges and how they are to be drawn (e.g. dashed, dotted, etc.). One can specify the anchor points where edges may connect to nodes. and whether the graph's orientation should be left-to-right or top-to-bottom.

- DAG [GNV88], also allows a range of specifications for a graph. A minimum specification is a list of edges. A detailed specification includes a list of edges with optional attribute-name/attribute-value pairs. The specification for the graph in figure 6.2 is:

```
draw nodes as Circle width .5 height .5;
draw LR_0 LR_3 LR_4 LR_8 as Doublecircle width .5 height .5;

draw edges pointsize 8;
LR_0 LR_2 label "SS(B)";
LR_0 LR_1 label "SS(S)";
LR_1 LR_3 label "S($end)";
LR_2 LR_6 label "SS(b)";
LR_2 LR_5 label "SS(a)";
LR_2 LR_4 label "S(A)";
LR_5 LR_7 label "S(b)";
LR_5 LR_5 label "S(a)";
LR_6 LR_6 label "S(b)";
LR_6 LR_5 label "S(a)";
LR_7 LR_8 label "S(b)";
```

```
LR_7 LR_5 label "S(a)";
LR_8 LR_6 label "S(b)";
LR_8 LR_5 label "S(a)";
```

The first line specifies that, by default, nodes should be drawn as a half-inch wide circle. The next two lines specify that a particular set of nodes (those representing final states) be drawn as double circles and that the edge labels should be drawn in pointsize 8. The remaining lines specify the list of edges and their labels.

The set of attributes available for nodes include shape (Box, Circle, Doublecircle, Diamond, etc.), size, label, color, pointsize, and rank. The set of attributes available for edges include label, pointsize, color, inkstyle (dashed, dotted, etc.), and weight. The weight is used to influence the layout of the graph because the layout algorithm will attempt to minimize the length of "heavy" edges. One can specify explicitly that an edge is a back edge (i.e. that its default orientation is opposite that of other edges).

6.1.3   External Formats for Other Tools

The external representation of the interactive tools Tweedle [Ase87], Lilac [Bro88] and Lefty [DK91] are relevant because they use a textual representation and an interactive representation simultaneously. The user can modify either one, and the system will automatically update the other representation to keep the two views consistent. The textual representation contains all information about the appearance and current state of the tool, and thus could be used to achieve persistence.

- Tweedle is an interactive drawing program. The user always has access to both a graphical and a textual representation of the drawing. The textual representation is somewhat similar to the PostScript [Sys85] page representation language used by many laser printers.

- Lilac is a document editor. The user has access to both an interactive WYSIWYG ("what you see is what you get") document editor and a textual representation containing the text and the formatting commands. The textual representation is a hierarchically-structured description of the document.

- Lefty is an editor for technical pictures. The textual representation is a program containing functions to draw and edit the picture. Lefty can communicate with other processes via standard inter-process communication facilities. For example, Lefty is used to provide an interactive front-end to the DAG graph drawing program.

Figure 6.3: IDL overview (from [Lam87])

As is also the case with an interactive graph editor, there are times when it is appropriate for the user to edit the graphical representation interactively and there are times when it is appropriate to edit the textual representation (e.g. when making global substitutions or some other repetitive task).

Eiffel

Eiffel [Mey87, Mey88] is an object-oriented language intended for the development of extendible and reusable software. One of the classes available in Eiffel is the "storage" class which defines two methods "store" and "retrieve" for converting a data structure to and from its external representation in a file. Any object which inherits from the class storage can use these methods. Eiffel is relevant because the proposed solution also achieves persistence via methods which save and load the object to and from a file.

When the store method is invoked, the object's data structure, including all objects it points to (whether they are members of the "storage" class or not) will be saved in a machine-dependent, binary representation. The data structure can be loaded by the same or another program as long as the class definition does not change.

Lisp

In Lisp, both the program and data are represented as list-structures and this makes it easy to save a representation of a data structure in a file.

IDL

IDL (Interface Description Language) [Sca89, Lam87, NNGS90] is a language for describing data structures that is used for sharing data structures between programs which may be written in different languages or running on different machines. The external representation used to communicate the data structure between tools is based on a textual flattening of the data structure.

```
structure simple Root exp Is
  tree ::= leaf | inner;
  inner => op: operator,
    depth: Integer,
    left: tree,
    right: tree;
  leaf => depth: Integer,
    val: Rational;
  operator ::= plus|minus|times|divide;
End
```

(a)

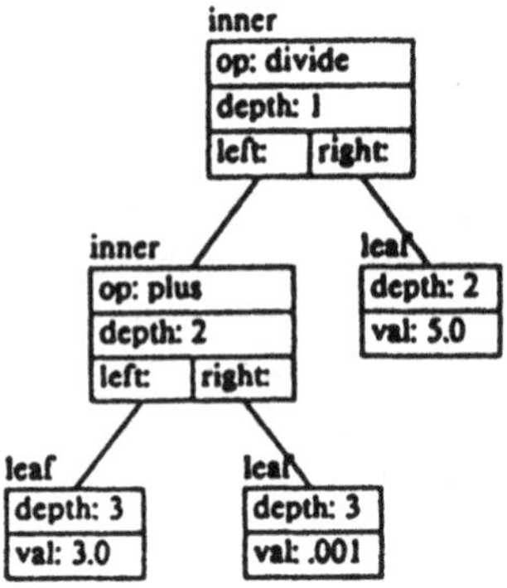

(b)

Nested External Representation	Flat External Representation
inner[op divide; depth 1;	L1: inner [op divide; depth 1; left L2↑; right L3↑]
left inner [op plus; depth 2;	L2: inner [op plus; depth 2; left L4↑; right L5↑]
left leaf [val 0.3E1; depth 3];	L4: leaf [val 0.3E1; depth 3]
right leaf [val 1/1000; depth 3]];	L5: leaf [val 1/1000; depth 3]
right leaf [val 5.0; depth 2]]	L3: leaf [val 5.0; depth 2]

(c)

Figure 6.4: IDL example (from [Lam87])

As shown in figure 6.3, a user's *process* in IDL consists of a reader, a writer, and an interface, all of which can be generated automatically from the IDL data structure description. The *reader* converts an external representation to an internal representation. The *writer* converts the internal representation to an external representation. Two external representations are available – nested and flat. Both forms of the external representation are generated automatically from the IDL description of the data structure by the IDL translator tool. The IDL description itself, however, must be written by the application developer.

In IDL, the external representation is used to pass data structures between a set of related tools. However, the same format could be simply written to or loaded from a file to achieve persistence. IDL is relevant because the format of the flat external representation of the data structure is similar to that of the proposed solution.

Figure 6.4(a) shows the IDL description for a tree data structure. This description is language-independent and must be provided by the user or application developer. Figure 6.4(b) shows the data structure of a such a tree for the expression "$-(3.0 + .001)/5.0$". Figure 6.4(c) shows the corresponding nested and flat external representations.

6.2 GRL: An External Representation for Graph- based Tools

The solution proposed here is based on the textual flattening approach to persistence. There are two reasons for chosing this approach. Firstly, a textual representation allows the user to edit the graph information easily. As was shown in section 6.1.3, there are times when it is appropriate to interactively make changes to a structure (e.g. using an interactive graph editor) and there are times when a textual interface is more appropriate. Changing all of the node and edge labels to uppercase or making a global substitution are two examples of editing operations which, although possible with a graph editor, are much simpler using a text editor. Secondly, assuming that incomplete entries are initialized to reasonable default values, an external representation with the aforementioned properties can be used to support a range of specifications. A minimal specification would consist simply of a list of nodes and edges. A full specification can be used to preserve the state of the graph editing session. Typically, the user would use a minimal specification when starting the editor the first time and then use the full specification subsequently.

This section presents GRL (Graph Representation Language), a language designed to be used as an external representation of an interactive graph editing tool [New88]. This section describes the standard portion of GRL. Appendix A describes the lexical conventions used and gives an extended BNF grammar of the GRL language. The next section describes how GRL can be extended so that application-specific attributes can also be specified. The extendibility of GRL is one of the novel aspects of this work. In contrast to IDL where the application developer has to provide an IDL description of the data structure, GRL allows a program generator tool to generate "readers", "writers", and even "editors" for the data structure automatically from the source code of the data structure declarations. In section 7.3 of the following chapter, the program generator tool is described in detail.

The input format of GRL is a text file containing a list of entries representing either graph, node, edge, or layout constraint information. Each entry contains the name of the data type followed by a list of attribute-name/attribute-value pairs as shown below:

$$<data\ type\ name>\{(<attribute\ name>:<attribute\ value>)^*\}$$

The order of the attribute-name:attribute-value pairs is irrelevant. The order of the entries themselves is also irrelevant with the exception that non-default entries should be defined before they are used. If an attribute is redefined, the newer definition takes affect. Comments in C or in C++ style may be present anywhere in the input. The list of attributes are based on those used by the EDGE graph editor (described in chapter 8), but they could easily be extended or modified for use by any other graph-based tool.

Default attribute values for node or edge types can be specified as part of the graph entry using a specification of the form:

$$<data\ type\ name>.<attribute\ name>:<attribute\ value>$$

The final example in section 6.4 shows an example of this usage.

If the editing session contains several views of a graph (see section 5.2.1), then a GRL description for each view will be saved in a separate file. The GRL file for the main graph editing session will specify the names of these files using the graph's include attribute.

Table 6.1 shows the list of attributes that can be specified for a graph object. Table 6.2 shows the list of attributes that can be specified for a node object. The minimal entry is the attribute "title". Table 6.3 shows the list of attributes that can be specified for an edge object. The minimal entry is the pair of attributes for "sourcename" and "targetname". Table 6.4 shows the list of attributes that can be specified for a layout constraint entry. In addition to the attribute "nodes", the minimal entry will include either the name of a single-dimension constraint together with a dimension specifier or the name of a multi-dimension constraint.

Attribute Name	Attribute Usage
title	specifies the name associated with the graph
x	the x-coordinate of the editing window
y	the y-coordinate of the editing window
width	the width of the editing window
height	the height of the editing window
xmax	width of the scrollable virtual window
ymax	height of the scrollable virtual window
xscrollbar	position of the horizontal scrollbar
yscrollbar	position of the vertical scrollbar
xbase	horizontal offset between editing window and graph
ybase	vertical offset between editing window and graph
xspace	horizontal spacing between nodes
yspace	vertical spacing between nodes
sspace	third-dimension spacing between nodes
color	background color of the editing window
bordercolor	border color of the editing window
topsort	which topological sorting algorithm to use
orientation	graph's preferred orientation (e.g. left-to-right, top-to-bottom, etc.)
layoutalgorithm	which layout algorithm to use
layoutfrequency	how often the layout should be performed (e.g. after every change or upon user request)
scaling	a scaling factor for the entire graph
inputfunction	the name of an application-specific input function
outputfunction	the name of an application-specific output function
sel_nodes	the set of currently selected nodes
abstraction_view	how should this subgraph be viewed (e.g. black-box, white-box, grey-box, separate view)
include	the name of an additional GRL file

Table 6.1: Standard set of graph attributes

Attribute Name	Attribute Usage
title	unique title of this node (Title may contain newline characters to allow multi-line entries.)
label	label shown in node's window
fontname	font used to display label
textmode	justification of text (left, right, or center)
width	width of node (by default max width of text and icon)
height	height of node (by default max height of text and icon)
borderwidth	width of node's window
color	background color of the node's window
bordercolor	color of the node window's border
iconfile	name of file containing a bitmap image
iconstyle	how to display the icon (e.g. surrounding the node or as part of the node)
ports	set of points where edge can connect to node

Table 6.2: Standard set of node attributes

Attribute Name	Attribute Usage
sourcename	title of the source node
targetname	title of the target node
label	label to be displayed on this edge
fontname	font used to display label
linestyle	how edge should be drawn (e.g. continuous, dotted, dashed)
thickness	thickness of the edge
color	color of the edge
arrowstyle	how the arrowhead should be drawn (e.g. none, line, solid)
arrowpos	position of the arrowhead (e.g. source, target, both)
arrowwidth	width of the arrowhead
arrowheight	height of the arrowhead
arrowcolor	color of the arrowhead

Table 6.3: Standard set of edge attributes

Attribute Name	Attribute Usage
name	Single-dimension constraints: equal, smaller, greater, neighbors, low_margin, high_margin, range, cluster or limit. Multi-dimension constraints: above, below, left, right, in_front, behind, equal_row, equal_column, top_margin, bottom_margin, left_margin, right_margin, upper_neighbor, lower_neighbor, left_neighbor, or right_neighbor.
nodes	list of nodes affected by constraint
dimension	x, y, or z dimension
priority	integer priority representing importance of satisfying this constraint
interval	two integers specifying the range of positions the nodes should lie within
title	the name of a cluster
size	the maximum size of a cluster

Table 6.4: Standard set of layout constraint attributes

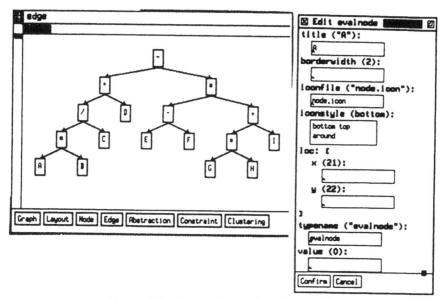

Figure 6.5: Expression evaluation graph

6.3 Language Extension

There are many applications in which additional application-specific attributes
will be used. Consider the application shown in figure 6.5 which displays and
evaluates integer expressions. Each of the nodes may take on a value. The user
assigns initial values to the leaf nodes in the input file or interactively through the
node's menu. When the expression is evaluated, the calculations are performed
and are passed up to the root of the graph where the expression's value is kept.
The user can examine each node's value through the node's menu.

The standard version of GRL described thus far can be used to specify the
graph representing the expression tree. Ideally, the user would also like to use
GRL to specify the additional application-specific value in the same way as the
standard attributes are set. For example:

```
graph: {
 evalnode: { title: "A"  value: 1}
 evalnode: { title: "B"  value: 2}
 ...
 evalnode: { title: "I"  value: 0}
}
```

The design of GRL is such that this facility is available with practically no
effort on the part of the application developer or user. This is accomplished by
using a program generator tool which generates the GRL scanner and parser from

the application-specific C++ class declarations. In addition, the tool can also
generate menus which include the additional attribute (as shown in figure 6.5).
This program generator tool which automates the extendibility of the graph
editor's external interface will be described further in chapter 7.

Although the standard version of GRL described thus far uses only simple
data types (integer, real, char, string, and enumerated types), the grammar of
GRL allows the specification of any C++ data type (including structures and
arrays) except pointers. The saving and restoring of pointers is beyond the scope
of GRL's design, but a solution is presented in [PMT90].

6.4   Examples

Simple Example

The following is a minimal GRL description for the graph shown on the right.
Normally, the GRL description contains a list of nodes as well as a list of edges,
but node descriptions may be omitted if the default values are sufficient.

```
graph: {
  edge: { sourcename: "A"  targetname: "B" }
  edge: { sourcename: "A"  targetname: "C" }
  edge: { sourcename: "C"  targetname: "D" }
  edge: { sourcename: "D"  targetname: "E" }
}
```

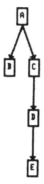

Abstraction Example

This example shows the same graph as in the previous example, but with one subgraph abstraction and with a different graph orientation.

```
graph: {
  node.fontname: "9x15bold"
  node.bordercolor: "white"
  orientation: left_to_right
  /* list of nodes */
  node: { title: "A" }
  node: { title: "E" }
  graph: {
    title: "sub1"
    /* a white boxed view of the abstraction */
    abstraction_view: white
    /* list of nodes */
    node: { title: "B" }
    node: { title: "C" }
    node: { title: "D" }
    /* list of edges */
    edge: { sourcename: "C"  targetname: "D" }
  }
  /* list of edges */
  edge: { sourcename: "A"  targetname: "B" }
  edge: { sourcename: "A"  targetname: "C" }
  edge: { sourcename: "D"  targetname: "E" }
}
```

Complex Example

This shows a more complex example. It demonstrates a hierarchical subgraph abstraction, different node types, different arrow styles, node icons, and a constraint specification. The layout algorithm is set to the "constraint" layout algorithm and the default font is set to bold helvetica font in point size 10. The next three lines specify default values for the attributes of three application-specific node types thick, high, and wide. This is followed by the list of nodes, the list of edges, and the list of constraints. The list of nodes contains six "normal" nodes and the seventh is a node representing a hierarchical subgraph abstraction. To demonstrate how easily the user can customize the appearance of nodes and edges, various node and edge attributes are set. A layout constraint is used to align the nodes a, b, i, and j.

```
graph: {
  layoutalgorithm: constraint
  node.fontname: "helvi0b"
  thick.borderwidth: 10
  high.height: 50
  wide.width: 50
  /* list of nodes of main graph */
  node: { title: "a"  iconfile: "target.icon" }
  wide: { title: "h"  textmode: right_justify }
  wide: { title: "k"  iconfile: "ellipse.icon"
          iconstyle: around }
  high: { title: "i" }
  high: { title: "j"  width: 30  borderwidth: 5 }
  thick: { title: "b" }
  graph: {   // outer abstraction
    title: "sub1"
    abstraction_view: white
    /* list of nodes of first subgraph */
    wide: { title: "d" label: "This is a
three-line
node" }
    high: { title: "c" }
    graph: {   // inner abstraction
      title: "sub2"
      abstraction_view: black
      /* list of nodes of second subgraph */
      node: { title: "f" }
      node: { title: "e" }
      node: { title: "g" }
      /* list of edges of second subgraph */
      edge: { sourcename: "e"  targetname: "f" }
      edge: { sourcename: "e"  targetname: "g" }
    }
    /* list of edges of first subgraph */
    edge: { sourcename: "c"  targetname: "d"  thickness: 5  arrowwidth: 5
            arrowheight: 10 }
    edge: { sourcename: "c"  targetname: "e" }
  }
  /* list of edges of main graph */
  edge: { sourcename: "a"  targetname: "b" }
  edge: { sourcename: "a"  targetname: "c"  arrowstyle: none }
  edge: { sourcename: "b"  targetname: "h"  linestyle: dashed }
  edge: { sourcename: "b"  targetname: "i"  arrowstyle: line }
  edge: { sourcename: "i"  targetname: "j"  linestyle: dotted
          arrowstyle: solid }
  edge: { sourcename: "h"  targetname: "k"  arrowheight: 10  arrowwidth: 10 }
  constraint: { name: equal_column  nodes: {"a"  "b"  "i"  "j"} } //constraint
}
```

6.5 Results

The flexibility of GRL allows a range of specifications from a simple format containing simply a list of nodes and edges to a format containing all details of the graph's structure and appearance as well as the current state of the editing session. The attributes can be specified in any order and the "include" command makes the GRL external representation modular. The language extension described in section 6.3 allows application-specific attributes to be specified in the same way as the standard set of attributes. Because GRL is a textual description of the graph this also allows the user to edit the GRL description to make changes to the graph. Thus, GRL satisfies all of the goals for an external representation given at the beginning of this chapter.

One of the main advantages of the solution to persistence proposed in this chapter is that it can easily and automatically be extended to handle application-specific attributes. This aspect will be presented in the following chapter.

Chapter 7

"To achieve reusability and extendibility, the principles of object-oriented design seem to provide the best known technical answer to date."
— B. Meyer

Extendibility

A program that is easy to adapt to a wide variety of applications is called *extendible*. Extendibility includes, but is not restricted to, customizability and extensibility which are defined as follows. A *customizable* program is one in which the user can specify attributes for existing features. An *extensible* program is one in which the user is able to add new features to a system in order to adapt the tool to accommodate unforseen situations. Often an extendible system is also a highly *reusable* system, meaning that a large amount of the software can be reused across applications.

This chapter shows that object-oriented design is a natural way to achieve extendibility in a graph editor. A program generator tool will be presented in this chapter which partially automates the extension of the graph editor to a particular application.

7.1   Related Work

The traditional, process-oriented approach to programming and the object-oriented approach to programming differ in their interpretation of the statement:

The essence of programming is performing operations on data.

The process-oriented approach puts emphasis on the **operations** whereas the object-oriented approach emphasizes the **data** [Mey88].

One of the basic tenets of software engineering is that software will change. Change to a system most often takes the form of change in the operations to be performed rather than a change in the data. For example change to a text formatting program is likely to take the form of additional formatting commands or changes to take an additional sort of output device into account. Changes to the data are, of course, also possible, but usually they are not drastic enough to pass beyond the abstract representation of the data.

Traditional program development approaches such as top-down functional design again put the emphasis on the operations. Initially, the program is viewed as a single operation which is then split into several sub-operations using a technique called stepwise-refinement. As argued in [Mey88], this approach to software development is not well suited for change or reusability. Consider, for example, the development of an interactive text editor. A process-oriented version of the aforementioned text formatting program would be so different from the interactive text editor that it would probably be easier to design and implement the program from scratch than to try to reuse parts of the design or implementation. This is because the top-down design is so strongly linked to the batch-oriented version of the program. An object-oriented version of the text formatting program, on the other hand, could relatively easily be extended to an interactive text editor form because the operations are associated with the objects and those do not change.

Object-oriented programming is not a new concept. Its roots go back (at least) to Simula 67, a programming language originally designed to support simulation applications. An object-oriented approach is particularly suited for simulation because the real-world objects being simulated could be directly matched with their representations in the program. The term "object-oriented" has become very popular in recent years and the fashion is to call many things "object-oriented" even though there is still some disagreement as to exactly what this term means. The basic definition of designing with the emphasis on objects is one that everyone can at least agree on. Most proponents of object-oriented design [Mey87, Str86, Seb89, Cox86, Gol84], however, agree on a more specific definition – that object-oriented design includes the concepts of data abstraction and inheritance.

- **Data abstraction:** An *abstract data type* is a user-defined type such that there is a strict separation between the data type's specification and its implementation. An unrelated procedure can access the data type only through the set of operations given in the specification. The implementation is kept hidden from the application [Par72]. The terms *private* and *public* are used in C++ [Str86] terminology to distinguish between the data structure and methods that are known only within the class and those which are known outside of the class.

In object-oriented programming, abstract data types are used to represent the objects. This contributes to extensibility and reusability because the implementation of an object may be changed without affecting either the specification or any other code which uses the object.

- **Inheritance**: Objects may be structured in hierarchical form such that each object inherits attributes from objects above it in the hierarchy. This mechanism allows the programmer to specialize an object by creating a new object that inherits from an existing object. This mechanism also allows a programmer to "factor out" commonalities among a set of objects. *Single inheritance* means that an object may only inherit attributes from one other object whereas *multiple inheritance* allows an object to inherit attributes from several other objects. When an object of class B inherits from class A, then the object is said to be *derived from* class A (A is called the *superclass* and B is called the *subclass*).

Inheritance contributes to extensibility and reusability because one can define a class in terms of the differences relative to an existing class rather than implementing an entirely new class.

Most object-oriented systems support *polymorphism* and *dynamic binding*, which means that an object may be referenced by different type names and operations can have different implementations depending on the type of its operand. The type of an object and which operation to perform on it is determined by the system at run-time. This allows a subclass object to be assigned to one of its superclass objects. Operations may be redefined for the subclass object, and these are called *virtual functions* in C++ terminology. The object can thus be affected by operations associated with the superclass or by operations which were redefined in the subclass.

In summary, the following definitions will be used: An *object* is an encapsulation of data and operations on that data. An object often corresponds to some real-world entity. A *class* is a description of a set of similar objects and these objects are said to have the same *type*. A class description is split into two parts – a description of the interface to the object and the implementation of the object. When inheritance is used, the terms *superclass* and *subclass* are used to denote the class from which or to which inheritance is performed. A *method* is an operation associated with a class. An *instance* is a particular instanciation of a class. For example, when using a class representing the set of students one might state that "John is an instance of the class student.".

7.2   An Object-Oriented Graph Editor

Recall that the main motivation for designing a graph editor was that graphs can be used to represent relationships in a wide variety of applications. In these applications the nodes typically represent some real-world object (or abstraction thereof) and the edges represent some relationship between these objects. This alone is adequate reason to design a graph editor as an object-oriented system. Additionally, the editor should be extendible. The application developer will want to add application-specific code. Therefore, it is desirable to provide the application developer with a clear and consistent interface rather than allowing the changes to be spread throughout the code and this is accomplished easily with object-oriented design.

This section presents an overview of the object-oriented design of an extendible graph editor. A graph consists of a set of nodes and a set of edges. The object-oriented graph editor provides class definitions for the objects graph, node, and edge. The application developer can define subclasses of the graph, node, and/or edge classes to be used in the application. All operations are defined as virtual functions, so that the application developer only has to provide code that differs from the default operations.

Graph, node, and edge objects have a set of display attributes which control their appearance. No matter how extensive this list of display attributes is, it can never be complete – there will always be applications that want to use additional attributes. Therefore, it should be easy to add such attributes and, ideally, the additional attributes should be treated the same as attributes in the standard set. Because the list of attributes is long and likely to be sparse, it is awkward to have to specify them as part of a long, fixed-length list of parameters. Instead, the interface used for the display routines will pass a single parameter for the display attributes. This parameter will be a null-terminated list of pairs where the pairs will contain the attribute name and the attribute value.

7.2.1   Graph Class

Data Structure

The data structure for a graph includes a set[1] of nodes and a set of edges. The title of a node is used as the unique identifier for a node. For an edge, the source node and the target node are used together to form the unique identifier. The graph data structure will include a set of nodes which are currently selected[2]. Furthermore, the graph data structure includes attributes concerning the display of the graph.

---

[1] A "set" is a class which defines a list of objects such that duplicate entries are not allowed. The "set" class offers a standard set of methods for adding and deleting members, accessing each member in turn etc.. It will not be described here further.

[2] An edge is selected by selecting its source and target node. Therefore a separate list of selected edges is not necessary.

Methods

new graph() creates a new instance of an empty graph of the default graph type. The display attributes will be initialized to a set of default values.

new graph(char* typename) creates a new instance of an empty graph with the specified graph type. The display attributes will be initialized to a set of default values for this type.

delete() deletes an instance of graph, freeing storage that was associated with this instance.

read(char* filename) reads an external representation of this graph from the specified file.

write(char* filename) writes an external representation of this graph to the specified file.

add_node(node* n) adds the specified node to the set of nodes.

delete_node(node* n) deletes the specified node from the set of nodes.

add_edge(edge* e) adds the specified edge to the set of edges.

delete_edge(edge* e) deletes the specified edge from the set of edges.

find_node(char* title) finds the node uniquely identified by this title in the set of nodes.

find_edge(node* s, node* t) finds the edge uniquely identified by this source and target node in the set of edges.

set_attributes(attr_list* al) sets the attributes specified in the list. The attributes are specified as a null-terminated list of attribute-name/attribute-value pairs.

get_attributes(attr_list* al) similarly, this retrieves the current set of attributes.

layout() performs a new layout of the graph using the current layout algorithm.

draw() displays the graph.

new_selection() reinitializes the list of currently selected nodes.

add_selection(node* n) adds the specified node to the list of currently selected nodes.

delete_selection(node* n) deletes the specified node from the list of currently selected nodes.

menu() displays a popup menu for editing the graph's attributes.

7.2.2 Node Class

Data Structure

A node will be implemented as a window[3]. The window system will detect which node the pointing device is pointing to when selection occurs. The class node consists of a reference to the window used to display the node, a reference to the graph in which this node is currently contained, a set of attributes controlling the display of the node, and several less-interesting, private variables used, for example, to mark nodes when detecting cycles etc. For efficiency reasons, the node data structure also contains references to linked lists of successor and predecessor nodes.

Methods

new node() creates a new instance of an empty node of the default node type. The display attributes will be initialized to a set of default values.

new node(char* typename) creates a new instance of an empty node of the specified node type. The display attributes will be initialized to a set of default values for this node type.

delete() deletes an instance of the node, freeing storage that was associated with this instance.

read(char* filename) reads an external representation of this node from the specified file.

write(char* filename) writes an external representation of this node to the specified file.

set_attributes(attr_list* al) sets the attributes specified in the list. The attributes are specified as a null-terminated list of attribute-name/attribute-value pairs.

get_attributes(attr_list* al) similarly, this retrieves the current set of attributes.

move(int x, int y) moves node to this coordinate.

draw() displays the node.

menu() displays a popup menu for editing the node's attributes.

---

[3]The assumption is made that the graph editor will be implemented on a system, such as the X Window System, in which windows are a cheap resource.

7.2.3 Edge Class

Data Structure

The class edge consists of a reference to the graph in which this edge is currently contained, references to the source and target nodes of the edge, and a set of attributes controlling the display of the edge.

Methods

new edge() creates a new instance of an empty edge of the default edge type. The display attributes will be initialized to a set of default values.

new edge(char* typename) creates a new instance of an empty edge of the specified edge type. The display attributes will be initialized to a set of default values for this edge type.

delete edge deletes an instance of the edge, freeing storage that was associated with this instance.

read(char* filename) reads an external representation of this edge from the specified file.

write(char* filename) writes an external representation of this edge to the specified file.

set_attributes(attr_list* al) sets the attributes specified in the list. The attributes are specified as a null-terminated list of attribute-name/attribute-value pairs.

get_attributes(attr_list* al) similarly, this retrieves the current set of attributes.

draw() displays the edge.

menu() displays a popup menu for editing the edge's attributes.

7.3 AGENT: A Tool to Automate Extendibility

Although the use of object-oriented techniques saves the application developer a considerable amount of development time, there is room for improvement. Much of the code provided by the application developer is repetitious and could be produced automatically by a program generator tool instead.

One of the design goals of GRL is to be able to parse application-specific attributes as easily as the standard set of attributes. This raises the question of how to extend the GRL scanner and parser so that the application-specific attributes can be included with a minimum of effort on the part of the application developer.

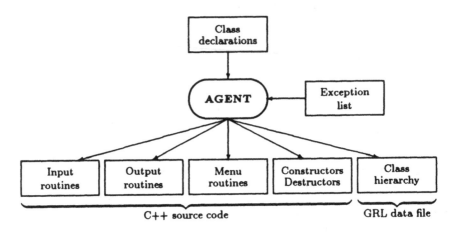

Figure 7.1: AGENT program generator tool

The proposed solution uses a program generator tool called AGENT (**A GEN**erator **T**ool) [Man90a, PM90] to assist in the generation of the source files. Program generation is a technique that is becoming increasingly popular and that has been used successfully in data processing, data bases, user interfaces, and parsers. Program generators translate specifications into various forms – typically segments of program code. The specification is usually a high-level description of the tasks the generated output should implement. This technique is practical when it is significantly easier to change the input specification than to change the code itself. More information about program generation techniques can be found in [Cle88, LB86, vHE87].

Figure 7.1 shows an overview of the AGENT program generator tool. It reads the standard set of class declarations used by the graph editor as well as the class declarations written by the application developer. From these class declarations, the program generator tool extracts information regarding the class hierarchy, the set of node and edge classes, and the name and type of each attribute. Any C++ data type may be specified in the class declarations. From this information, the program generator tool produces a set of input, output, and menu routines. These routines will be able to read, write, and edit application-specific attributes in addition to the standard set of attributes. Source code for simple constructors and destructors for the classes are generated. A GRL file containing a description of the class hierarchy is also generated. The generation can be controlled by the user via the exception list to inhibit the generation of code for particular attributes or entire classes.

This section will describe the code generated by AGENT. Section 7.4 presents an example of how an application developer would use AGENT to automatically generate large portions of a logic simulator application.

7.3.1 Input Routines

The input routines are a set of procedures that read input in the GRL format and create node and edge instances. Although an earlier version of AGENT generated the appropriate lex [LS75] and yacc [Joh75] input files [Man90a], the current version generates C++ source files directly.

7.3.2 Output Routines

The output routines are a set of procedures that save the graph editing session. A hand-coded, fixed portion saves information associated with the graph and the editing session in GRL format. To save each node, the write procedure associated with each node is invoked. The code of the write procedure for each node or edge is generated by the program generator tool. This is a virtual function which will write out attributes of the node or edge in GRL format. The write procedure will check whether an attribute has the default value before writing it out. If so, then that attribute will not be written out. This is done to prevent the GRL file from getting unreasonably large. When the user saves the graph, application-specific fields will also be saved.

7.3.3 Menu Routines

The generated source code displays a menu containing each attribute. Figure 7.2 shows an example of a generated menu. This includes not only the standard set of attributes, but also the application-specific attributes. By relying on AGENT to generate the menus, an application developer can avoid the large amount of detail work normally associated with the implementation of the graphical user interface.

7.3.4 Constructors and Destructors

Constructor and destructor procedures are generated for each node and edge type specified in the C++ class declarations. These procedures are used to allocate and deallocate storage for the nodes and edges. The default constructor will allocate storage for the object and will set the default values for the attributes. The application-specific attributes are set explicitly in the constructor. The inherited attributes are set recursively by the invocation of the constructor for the default node.

7.3.5   Class Hierarchy

AGENT generates a GRL file describing the class hierarchy. This can be used
to display the class hierarchy graphically. Figure 7.2 shows an example where
one window displays the class hierarchy while the main graph editing window
edits the application graph.

7.3.6   Exception List

AGENT generates procedures for the constructors, destructors and menu routines declared in the object's class declarations. By default AGENT generates
source code for all attributes in the specified classes. But often there are attributes which are intended for internal use only. Such attributes should not be
edited interactively or specified in a GRL data file. To prevent AGENT from
generating source code for these attributes, the application developer can specify
a list of the attributes for which no output should be generated. Similarly, the
application developer can specify that AGENT should ignore specified procedures or even entire classes. This exception list is read by AGENT at startup
time.

The syntax used to specify attributes in the exception list is simple. Attributes are specified as

<class name>.<attribute name>
or
<class name>.{<attribute name> (,<attribute name>)* }

for individual attributes or a list of attributes respectively. For example to ignore
attributes "dummy" and "visible" of class node the exception list could contain
either the entry:

`node.{dummy,visible}`

An entire class can be ignored by specifying the class name in the exception
list. The exception list can also be used to suppress the generation of particular
functions. The syntax for this is

[<class name> .] <function name> ()

For example the entries

```
node ()
~node ()
node.menu_show ()
menu_edit ()
```

in the exception list would cause AGENT not to generate a constructor, destructor and the menu_show function for class node and not to generate the menu_edit function for any class.

7.4 Example

Consider a VLSI application that displays a logical circuit and simulates its operation. Figure 7.2 shows an example of a circuit representing a two bit comparator. In addition to the main editing window containing the circuit, a separate view showing the class hierarchy as well as a menu for editing an input node are displayed. The nodes A0 and A1 represent the two bits of one input and the B0 and B1 nodes represent the two bits of the other input. In this circuit there are three output nodes LESS, EQUAL, and GREATER and one of them will have the value "1" depending on the input values. The edge labels represent the values propagated through the circuit. In figure 7.2 all inputs are zero so the edge leading to the output node EQUAL is labeled "1" (true). When the user selects one of the input nodes and requests the "edit node" operation, a menu will be presented in which all attributes of the node (including the value for this node) can be specified. After selecting "Confirm" in the menu, the circuit will be reevaluated and the graph will be redisplayed showing the new output values.

Different node types are used to denote the logical operations "and", "or", "not" and "exclusive or". Two further node types represent the input and output values. Each node type has an integer variable called value associated with it which will be used in this application to store the value of a node at this point in the circuit. The application developer provides procedures that evaluate the network, i.e. that propagate the values of the input nodes through the set of AND, OR, NOT, and XOR nodes performing the appropriate bit operations at each node.

A partial listing of the GRL input file for this application is shown in figure 7.3. Note that the typenames give in the GRL input file are identical to those in the class specification shown in figure 7.4.

To generate a complete set of constructors, destructors, input, output, and menu functions for this application, the application developer would invoke AGENT specifying the file names of all of the class declarations. To use the generated code, the application developer compiles it together with the other application-specific code and the rest of the graph editor to create a working application.

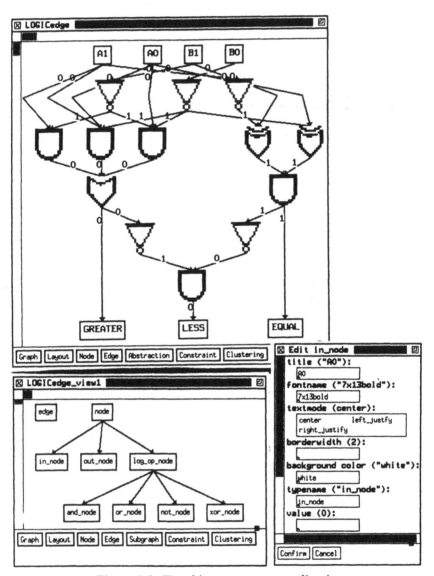

Figure 7.2: Two bit comparator application

```
graph: {
  /* default initialization for node types */
  and_node.iconfile: "../icon/and"
  or_node.iconfile:  "../icon/or"
  xor_node.iconfile: "../icon/xor"
  not_node.iconfile: "../icon/not"

  /* list of nodes */
  in_node:  { title: "A0"  value: 0 }
  in_node:  { title: "A1"  value: 0 }
  in_node:  { title: "B0"  value: 0 }
  in_node:  { title: "B1"  value: 0 }
  out_node: { title: "GREATER" }
  out_node: { title: "EQUAL" }
  out_node: { title: "LESS" }
  not_node: { title: "N1" }
  and_node: { title: "AND1" }

  ...

  /* list of edges */
  edge: { sourcename: "A0" targetname: "AND1" }
  edge: { sourcename: "B1" targetname: "N1" }
  ...
}
```

Figure 7.3: GRL input for two bit comparator application

7.5   Results

Object-oriented design of software is one of the best currently available techniques to keep the cost of software development down. This is because the
concepts of data abstraction and inheritance allow the application developer
to concentrate on the application. Data abstraction provides a clear separation between the specification and the implementation of objects. Inheritance
allows the application developer to define a new application-specific object in
terms of existing basic objects. The application developer can thus concentrate
on the application-specific attributes without necessarily understanding details
provided by the base classes. Automation can further relieve the application
developer from the "busy work" of extending an object-oriented system.

   The generated code takes advantage of the object-oriented nature of the
system by generating standalone components for the base classes whereas components for the derived classes consist of an invocation of the component for the
base class together with specialized code for the derived class. One could argue

```
class in_node : public node {
  int value;
public:
  in_node (char *);        // constructor
  ~in_node ();             // destructor
  void read ();            // reader
  void write ();           // writer
  void menu_add ();        // menu for adding
  void menu_edit ();       // menu for editing
  void menu_show ();       // menu for showing
};

class log_op_node : public node {
...
};

class and_node : public log_op_node {
...
};

...
```

Figure 7.4: Class declarations for two bit comparator application

that since the code is automatically generated anyway, there is no advantage to generate it in object-oriented style. Generating it in this way, however, makes it easier for a human to use or modify. It is clearer and more concise. It is conceivable that the application developer may make changes to a class but still want to derive all the others. By relying on generated code the application can at least get a prototype version of the application running quickly.

This chapter shows that a program generator tool can be used to relieve the application developer of a large part of the customization effort. A basic set of input, output, and editing routines, which the user can either use directly or further extend, are generated automatically. Most program generators rely on the user to provide a separate, high-level description which is used only by the program generator. The generation of program code from the class declarations themselves, rather than from a separately developed specification, is a powerful and novel technique. In addition to saving programming effort, the resulting code is practically guaranteed to exactly match the specification given by the class declarations. This makes it simpler to add additional attributes without having to worry about keeping other pieces of the system up-to-date.

> "*Invention or discovery,*
> *be it in mathematics or*
> *anywhere else takes*
> *place by combining ideas.*"
> – J. Hadamard

EDGE: An Extendible Graph Editor

EDGE (**E**xtendible **D**irected **G**raph **E**ditor) is a graph editor which can be used for many different applications. Although originally designed for editing directed graphs, EDGE can also edit undirected graphs. After presenting an overview of EDGE, this chapter will concentrate on the facilities EDGE provides for graph layout, graphical abstraction, persistence, and extendibility. A set of examples of EDGE applications is given at the end of the chapter.

Figure 8.1 shows a sample EDGE session. A graph is displayed in the EDGE window. A set of pull-down menus is available to manipulate the graph. Horizontal and vertical scrolling facilities are available, thus allowing the user to edit large graphs. A layout overview ("MiniView") of the graph is displayed upon request by the user. All of the windows (the main EDGE window, the pop-up menus, the separate views etc.) may be manipulated (e.g. moved, resized, iconified etc.) by the user through X Windows System.

The graph description is given as a GRL file and is shown on the righthand side of the screen. Before loading a graph, EDGE reads a startup file which is also in GRL format. This startup file can be used to set default values which the user wants to use while viewing several different graphs.

The user can add, delete, edit, or show nodes or edges in the graph. Add and edit cause a pop-up menu to be displayed in which the user can enter values for the attributes. Show causes a window containing a read-only listing of the attributes to be displayed. The user can specify a set of nodes or edges

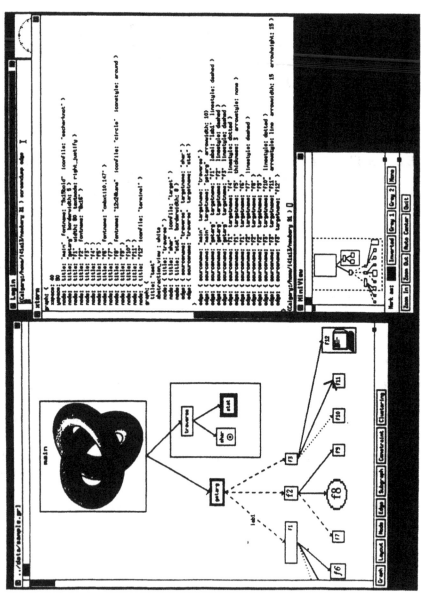

Figure 8.1: Terminal session showing the EDGE graph editor

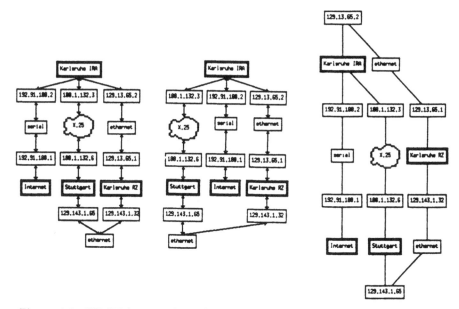

Figure 8.2: EDGE layout algorithms: Sugiyama, tree, undirected planar

by selecting them with a pointing device. The selected operation is always performed on the set of selected objects. Commonly used, simple operations, are also available as "keyboard accelerator" functions, meaning that the operation is performed when the keystroke sequence is entered rather than presenting the user with a menu.

Each node is implemented as an X-window. In the X Window System, windows are an inexpensive resource [SG86]. Hundreds, but not necessarily thousands, of windows can be displayed. It is not reasonable to display more than that anyway since the user cannot comprehend so much information at once. Abstraction mechanisms can be used to restrict the number of nodes displayed at any one time.

8.1 Graph Layout

EDGE is designed so that it may be used for many different graph-based applications. No currently available graph layout algorithm is appropriate for all applications. Therefore, EDGE offers several different layout algorithms and it is relatively easy to add additional ones. The following describes the set of layout algorithms currently available in EDGE. All are described in more detail in section 4.1. Figure 8.2 shows examples of the Sugiyama, tree, and undirected planar layout for a representation of part of our local network configuration.

Sugiyama: The Sugiyama algorithm [STT81] (see section 4.1.4) is appropriate for hierarchical directed graphs. This algorithm is one of the most commonly used for EDGE applications as well as for other systems requiring automatic graph layout. In addition to the standard Sugiyama algorithm EDGE offers a modified version of this algorithm which takes user- or application-specified layout constraints into account. The constraint layout algorithm is described in detail in section 4.2. Constraints can be specified either in the input file, by the application, or interactively by the user.

Tree: The tree layout is a linear time algorithm with the aesthetic goals of hierarchy, symmetry, and always displaying isomorphic subgraphs alike. It will also lay out graphs that are not trees by ignoring non-tree edges when positioning the nodes.

ISI: The ISI layout is a simple, linear time layout algorithm used by the ISI grapher [Rob87]. Leaf nodes are placed a fixed distance away from previously placed leaf nodes, and all other nodes are placed at the average position of their children. For the example graph given in figure 8.2 the ISI layout was identical to the tree layout. An example of the ISI layout on a more complex graph can be seen in figure 3.10.

Undirected Planar: The planar layout algorithm [Woo81] for undirected graphs is described in section 4.1.1. The implementation used in EDGE extends the algorithm in that trees which hang from the biconnected components are displayed as trees rather than as an arbitrary planar structure [Rum89].

8.2 Graphical Abstraction

In addition to the layout overview, EDGE provides a "focus" operation as a navigation aid. The user selects a graph object (either by name or by direct manipulation) and the system will automatically scroll the graph so that this object is centered in the editing window.

When viewing large or complex graphs, it is often useful to abstract away some of the detail so that important relations in the graph become clearer. EDGE offers several ways of simplifying the graph: hierarchical subgraph abstractions, separate views, and edge concentrations.

8.2.1 Subgraph Abstraction

EDGE offers an "abstraction" menu with operations for defining the subgraph and for switching between the different representations. A *subgraph abstraction* is a subgraph which can either be displayed in the context of the graph (using "black-box", "grey-box", or "white-box" representations as described in section 5.2.1) or in a separate editing session called a "separate view" (see section 5.2.1). The separate view starts a new graph editing session on the specified subgraph. There is no restriction on the number of separate views a user can have, and recursion of separate views is supported.

The subgraph can be defined using any of the techniques listed in section 5.2.2. The user can interactively add or delete nodes from an existing subgraph abstraction.

A subgraph abstraction can be "dissolved", which causes the nodes and edges within the abstraction to return to their parent graph. A "merge" menu is available to help the user merge changes made in the separate view back into the original graph. The merge command allows the user to specify whether the merger should take place as an "and", "or" or "copy" operation between the two graphs. The user can also specify whether the values of the attributes in the original graph or in the separate view should have priority. The matching between the nodes in the original graph and nodes in the separate view is currently done by comparing their title fields. This has the advantage of allowing the user to read in a GRL file in a separate view and then merge that graph back into the main graph. Extensions to EDGE which would allow the relationship between the nodes in the original graph and the separate view to be represented explicitly are currently under consideration.

8.2.2   Edge Concentration

An *edge concentration* is a set of directed edges such that all of the source nodes are connected to all of the target nodes. Edge concentrations were described in detail in section 5.3. EDGE offers a "clustering" menu with two operations: determine the set of edge concentrations, and concentrate the graph. To determine the set of edge concentrations, the user must specify a minimum concentration size and the level number to concentrate. When this operation is completed, EDGE will print out a list of the suggested edge concentrations. If the user is not satisfied with the results, the user may select a different size or level number and try again. Once satisfied with the suggested set of edge concentrations, the user selects the "concentrate graph" operation. This replaces each set of edge concentrations by their alternative representation as an edge concentration node. An undo facility, mentioned in "Future Work" (section 9.5), would be a useful feature, but this is not currently available.

8.3   Persistence

The default input/output format for EDGE is the graph representation language GRL which is described in section 6.2. Exactly the same set of attributes used in GRL are available in the menus. When the user selects the "save graph" operation, a menu is displayed in which the user can choose which attributes should be saved. Currently, the set of choices the user is given include: window information (size and scrolling position of the main editing window), layout information (layout algorithm and layout parameters), default values for object types, list of nodes, list of edges, and the list of constraints. This flexibility allows a user to save only the information which should be carried over to the next session. For example, users may choose not to save the window information because they would rather specify the size and position of the window when starting EDGE. The user can load a GRL description of a graph to start a new editing session. Similarly, the user can save and load subgraphs in GRL.

One of the attributes of GRL which may have been overlooked in chapter 6, is its ability to specify an alternative external representation. This allows application developers whose graph is not represented in GRL to use EDGE easily. To use this feature, the application developer must provide application-specific input and/or output procedures. The name of these procedures must be specified in GRL. EDGE will read the name of this procedure, look it up in its internal symbol table, and invoke this procedure on load and/or save operations. An examples of this technique will be presented in 8.6.2.

8.4 Extendibility

There are several different methods for adapting the EDGE editor kernel to a particular application. The choice of method depends on the extent of the desired customization. The following list summarizes the available methods. Any combination of these methods is permitted. In all cases, any application-specific code is loaded together with the rest of the EDGE system to form a single executable file.

8.4.1 Changing Visual Appearance

EDGE has a set of pre-defined attributes that control the appearance of the graph (e.g. the font name for node labels, the thickness of edges, etc.). If the graph display of an application needs a visual appearance that differs from the default only in the values of pre-defined attributes, then those values can simply be specified in the GRL external format. An example of this approach is the entity-relationship editor described in section 8.6.1.

8.4.2 Interfacing EDGE with an Existing Application

If the application developer wishes to add a graph display capability to an existing application with minimal changes to the application itself, the interfacing method may be chosen. After all, one should be able to reuse software developed prior to the existence of EDGE. In this method, the data structures of the application and of EDGE are separate, but "shadow" each other. This means that whenever one of the EDGE or application data structures changes, the other data structure must be updated as well. EDGE provides a well-defined interface for shadowing in the form of class definitions for graphs, nodes, and edges. Operations for the default graph class include: read, write, layout, constrain, and redisplay a graph. Operations for the default node class include: add, delete, edit, and draw as well as an action operation (used to perform some application-specific action). Operations for the default edge class include: add, delete, edit, draw and action. Each of these operations makes the appropriate changes in the EDGE data structures and subsequently invokes a corresponding "application interface procedure". These procedures are initially empty, but are meant to update the application data structures. These are the only procedures that must be extended for interfacing to EDGE. An example of this approach is the PERT chart editor described in section 8.6.3.

8.4.3 Integrating EDGE with a New Application

When constructing a new application, it is often simpler and more elegant to integrate application and EDGE data structures by extending EDGE's basic class definitions. The application may add new attributes and operations and even redefine existing operations. An extra benefit of this approach is that EDGE can now read and write the entire data structure from/to the file system, and provides a menu-controlled interface editing the data structures interactively. An example of this approach is the directory editor described in section 8.6.5.

8.5 Implementation

EDGE is written mostly in C++, with low-level routines (mostly the interface to the window system) written in C. The user interface of EDGE is based on the X Windows System. The reliance on these standard, widely available tools has made EDGE a very portable system. EDGE runs locally on Sun 3, Sun 4, and DEC 3100 UNIX-based workstations.

EDGE is roughly 30,000 lines of code. This includes approximately 17K lines of C++ code, 9K lines of C code, and 4K lines of generated C++ code.

The source code and binaries are available by anonymous ftp. For more information concerning the availability of EDGE, please send electronic mail to edge-request@ira.uka.de.

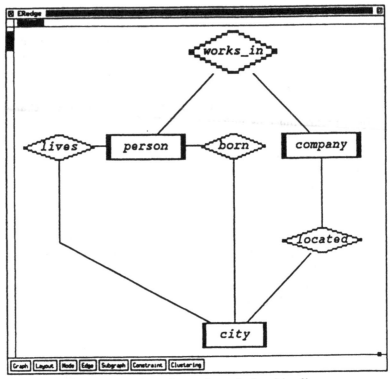

Figure 8.3: A browser for entity-relationship diagrams

8.6   Examples

To demonstrate the prevalence of graphs in applications and to show how easy it
is to extend EDGE, several application examples will be described. The examples
span the range of changing the visual appearance of EDGE, interfacing EDGE
with an application, and integrating EDGE with a new application.

8.6.1   A Browser for Entity-Relationship Diagrams

Figure 8.3 shows a browser for entity-relationship diagrams which is obtained
by simply customizing the visual appearance of EDGE. In this application, two
derived node types are used to display the entity-relationship diagram – one for
entities and one for relationships. Entities are displayed with a rectangular icon
and relations are displayed with a diamond-shaped icon and the node's label is
displayed inside the icon. EDGE automatically resizes the icon to fit around

the text (unless a fixed node size is specified). The planar layout for undirected graphs is used to layout the graph.

This application could be further extended by interfacing it to a database rather than relying on an external textual representation of the information (i.e. the GRL file).

As can be seen by looking closely at the relation nodes in Figure 8.3, EDGE always displays nodes within a rectangular area because a rectangular X-window is used to display nodes. The appearance could be improved by providing an application-specific "draw edge" function which draws through the node's window or by supporting non-rectangular windows.

Summary of the Application Developer's Effort

- provide a GRL file which set the attribute "iconstyle" to "around" and the "iconfile" attribute to the filename of a rectangular- or diamond-shaped bitmap depending on the node type.

8.6.2 A Tool for Visualizing Software Configurations

Figure 8.4 shows a tool for visualizing the structure of software configurations which is obtained by customizing the input procedure for EDGE. The graph shown in the top part of figure 8.4 is obtained by piping the output of the makedepend program (which generates a list of dependencies among source files) for a small C program into EDGE. The graph in the bottom part of figure 8.4 is obtained by using the "edge concentration" clustering technique.

Summary of the Application Developer's Effort

- provide an application-specific input procedure, e.g. "DependStyleRead", which takes the list of dependencies as generated by the makedepend program and invokes the corresponding "add node" and "add edge" operations.

- provide a startup file for EDGE which specifies the name of this application-specific input procedure. For example, create a startup file called ".dependstyle" containing:

      graph: inputfunction: "DependStyleRead"

- specify the argument "-" when starting EDGE to indicate that the input should be read from the standard input rather than from a file, e.g.:

      makedepend -f - *.c *.h | edge -style .dependstyle -

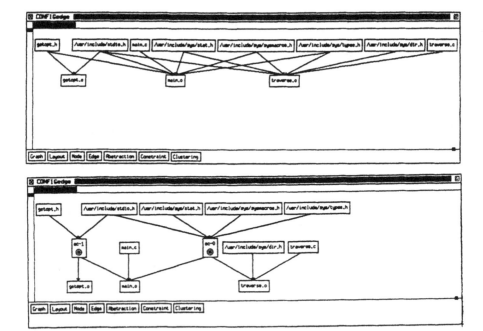

Figure 8.4: A tool for visualizing software configurations

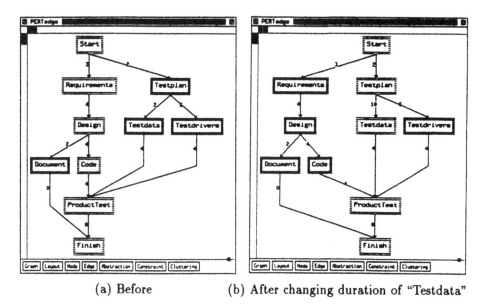

(a) Before   (b) After changing duration of "Testdata"

Figure 8.5: A PERT chart editor

8.6.3 A PERT Chart Editor

Figure 8.5 illustrates a PERT (Program Evaluation and Review Technique) chart editor built with EDGE using the interfacing technique (in which the application's data structures and EDGE's data structures are separate). A *PERT chart* is a graph that models the activities to be performed in a project[CCP87]. The nodes represent the activities and the edges specify the precedence relationship among the activities. The duration of an activity appears on its incoming edges. The *critical path* of a PERT chart is a path through the graph, where the sum of the activity durations is maximal. The critical path determines the minimal duration of the entire project, and any delay on that path causes the entire project to be delayed. Every time the graph is redrawn, the PERT chart editor highlights and aligns the nodes on the critical path.

Consider the PERT chart in figure 8.5 (a). When the user changes the duration of activity "Testdata" to 10, the critical path is recalculated and the set of nodes on the critical path are re-aligned as shown in figure 8.5(b).

Summary of the Application Developer's Effort

All of the changes necessary for this application involve "filling in" the (by default empty) set of application interface procedures provided by EDGE.

- provide the application code which manipulates its own data structure and calculates the critical path of the project. For clarity, all application-provided procedures used in this example are prefaced with the characters "CP_".

- change the application interface procedure APPL_DrawGraph to invoke an application-provided procedure which calculates the critical path.

- extend the application interface procedures APPL_AddNode, APPL_AddEdge, APPL_DeleteNode, APPL_DeleteEdge, APPL_EditNode, APPL_EditEdge so that they invoke an application-provided procedure which makes the appropriate changes to the application's data structure. For example:

```
node* APPL_AddNode(node* n) {
    CP_CreateNode(n->gettitle());
    return n;
}
```

- change the application interface procedure APPL_DrawNode to highlight nodes that are on the critical path. CP_IsOnCpath() is an application-provided function that checks whether the named node is on the critical path by looking up this information in the application's data structure. For example:

```
APPL_DrawNode(node* n) {
    if (CP_IsOnCpath(n->gettitle()))
        Set_Border(n->getwindow(), "grey");
    else
        Set_Border(n->getwindow(), "black");
}
```

- extend the application interface procedure APPL_Constraint so that nodes on the critical path are aligned. The previous set of alignment constraints (if any) are deactivated, a list of all nodes on the critical path is created, and this list is passed to the vertical alignment constraint procedure:

```
APPL_Constraints(graph* g) {
    node* n;
    /* undo previous constraints */
    if (connodelist != NULL) {
        g->constraints->equal_column(APPL, connodelist, UNDO, 0);
    }
    /* create node list */
    connodelist = new nlist;
    for (n = g->getheadnode(); n != NULL; n = n->getnextnode()) {
        if (CP_IsOnCpath(n->gettitle()))
            connodelist->append(n);
```

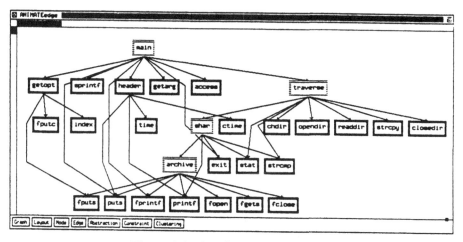

Figure 8.6: A call graph animator

```
}
/* give node list to straight line constraint */
g->constraints->equal_column(APPL, connodelist, DO, 0);
}
```

8.6.4 A Call Graph Animator

Figure 8.6 shows how an application that performs simple program animation can be integrated with EDGE. As the program executes, the active modules are highlighted so that the user can watch the program execute. Although one can hardly do justice to this in a static picture, figure 8.6 shows the screen partway through the execution of a program[1].

Summary of the Application Developer's Effort

- provide class declarations for a derived node which contains an application-specific field used to indicate whether the procedure this node represents is currently activated. The procedures Activate and Deactivate set this field. The AGENT program generator tool is used to generate a default constructor and destructor for this node type. The class declaration for "animatenode" is as follows:

```
class animatenode: public node {  // node is default node
    int active;
```

---

[1]The program being animated is the "shar" (shell archive) program which archives files so that they are extractable by the Bourne shell command interpreter.

```
public:
    animatenode(char*);     // constructor
    ~animatenode();         // destructor
    Activate()              {active = 1;}
    Deactivate()            {active = 0;}
    int IsActive()          {return (active == 1);}
};
```

- modify the source code to be animated by inserting calls to "Activate" and "Deactivate" at the beginning and end of each procedure to be animated.

- provide a description of the call graph of the program. The application developer can provide either a GRL description or use the output of a program that extracts the call graph from the source code together with an application-specific input procedure.

- extend the application interface procedure APPL_DrawNode to highlight the node if it is active. For example:

```
APPL_DrawNode(node* n) {
    if (n->IsActive())
        Set_Border(n->getwindow(), "grey");
    else
        Set_Border(n->getwindow(), "black");
}
```

- replace the call to "edit" in the procedure "main" of EDGE (which starts an infinite loop to handle the events received from the user) by a call to the (renamed) main procedure of the application program. This will cause the application to start executing after EDGE is initialized.

8.6.5   A Directory Editor

Figure 8.7 illustrates an editor for the UNIX directory tree integrated with EDGE. A directory editor displays the desired directory tree on the screen, permits graphical navigation in the tree, and allows the user to change the directory by editing the graph. The editor uses the tree layout algorithm of EDGE.

Three different node types are used in this application to represent directories, executable files, and non-executable files. Each node type is displayed differently. Directories have a border in background color, executables have a black border, and non-executables have a grey border. The size of the border is proportional to the size of the file. Also, each node type has different actions associated with it. Selecting the "action" operation for a directory causes that directory (and, recursively, all of its contents) to appear as a hierarchical subgraph abstraction. Figure 8.7 shows the appearance of the directory editor after

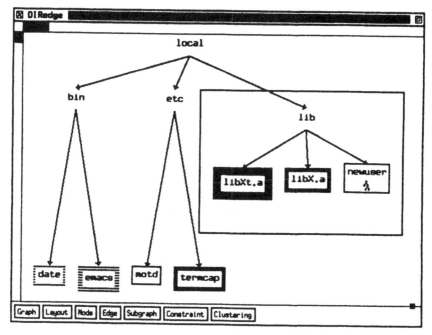

Figure 8.7: A directory editor

the subdirectories "lib" and "newuser" were selected and made into subgraph abstractions. Selecting the "action" operation for an executable node executes that file, while selecting a non-executable node causes an editing session to start on that file.

Selecting "delete" for any node will delete the node from the graph as well as delete the file from the file system. Selecting "add" for a directory node, will start an editing session on a new file that will be added to this directory. This directory editor could be further extended to interface with the command interpreter. Then the user could use the editor to specify files by selecting them with the pointing device rather than typing in their name. For example, to compile a file, the user could type the compile command in a command interpreter window and then select the file.

Summary of the Application Developer's Effort

- provide class declarations for the three node types. The AGENT program generator tool is used to generate constructors and destructors for these node types. The declaration for the node type for executables, for example, contains:

```
class xfilenode: public node {
```

```
class xfilenode: public node {
    char* machine_type;  // type of machine (e.g. Sun,...)
public:
    xfilenode(char*);    // constructor
    ~xfilenode();        // destructor
    draw();
    action();
};
```

- provide an input procedure that traverses the directory structure and invokes the "add node" (specifying the appropriate type) and "add edge" operations. The application interface procedure APPL_ReadGraph was extended to invoke this procedure. This could also have been accomplished by specifying the input procedure name in a startup file, but specifying it in the application interface procedure has the advantages: 1) that it is faster to startup because it is compiled into the code rather than using the dynamic lookup feature and 2) it allows the application developer to pass additional parameters to the input procedure which are used here to control the maximum depth of the directory.

- provide application-specific code for the "draw node" and "action node" operations for each node type. For example:

```
xfilenode::draw() {
    Set_Border(this->getwindow(), "grey");
}

xfilenode::action() {
    if (system(this->gettitle()) != NULL)
        error("Invoking command %s failed.\n", this->gettitle());
}
```

- extend the constructors generated by AGENT to allow for file creation and to set the borderwidth of the node proportionally to the file's size.

- extend the destructors generated by AGENT to delete the corresponding file when the node is deleted.

8.6.6   A Logic Simulator

The logic simulator, which was described in section 7.4, is another example of integrating EDGE with an application. Figure 8.8 shows a circuit representing a full one-bit adder. The nodes in the graph represent boolean operations which are performed on the input values to determine the output values. A typical use is that the user selects an input node, changes its value, and watches the calculated values propagate through the network to the output values. Different node types are used to represent each boolean operation.

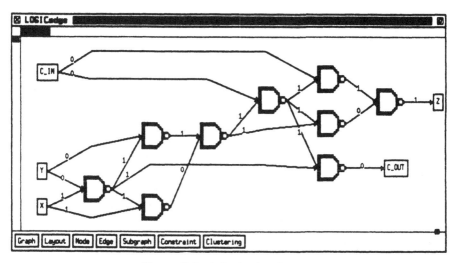

Figure 8.8: A logic simulator

Summary of the Application Developer's Effort

- provide class declarations for each of the node types. An example was shown in figure 7.4. The AGENT program generator tool is used to generate constructors and destructors for these node types. AGENT also generates menus containing the additional, application-specific "value" field.

- provide a GRL file describing the circuit. An example was shown in figure 7.3.

- provide an application procedure, e.g. called "evaluate_net", which propagates the values through the network. An excerpt is shown here:

```
evaluate_net(graph* g)
{
    ...
    if (strcmp(typename,"and") == 0) value = value && 1;
    else if (strcmp(typename,"or") == 0) value = value || 1;
    else if (strcmp(typename,"not") == 0) value = !(1);
    ...
}
```

- change the application interface procedure APPL_DrawGraph to invoke the procedure which evaluates the network.

8.7 Results

Table 8.1 summarizes the effort involved for each of these applications both in terms of lines of (C or C++) code as well as an estimate of the application developer's effort[2]. A distinction is made between interface code and application code. *Interface code* is code related to the interface between the graph editor and the application. This includes the declaration of derived node and edge types and the invocation of application-specific procedures. The *application code* is purely application-specific code.

Application	Lines of Interface Code	Lines of Application Code	Est. Effort in Person-Hours
ERedge	0	0	1
CONFIGedge	44	20	3
PERTedge	70	284	12
ANIMATEedge	29	101	8
DIRedge	55	192	8
LOGICedge	103	83	12

Table 8.1: Customization effort

The reader should bear in mind that this table shows the lines-of-code counts for "bare-bones" applications. Each of the applications described in this chapter provide only the first step towards a fully functional application. The motivation behind the design and development of EDGE is to show that EDGE can be easily adapted to a wide variety of applications, not to provide full-fledged applications.

In addition to being used within my own department, EDGE has been distributed to numerous universities, research centers, and companies both in Europe and the United States. The interest in EDGE is large and continues to grow. The list of applications for which "real" users use (or intend to use) EDGE include:

- **Program testing:** At the University of Bremen, EDGE is used as part of a tool for integration of testing of large software systems[HSF90].

- **Reverse engineering:** EDGE is used as part of the Maintainer's Assistant project at Siemens RTL to visualize the architecture of large software systems. One of the goals is to compare the original specification with the actual one in the code to detect inconsistencies. They have added facilities for automatically clustering the graph based on cross-reference information extracted from the software[SP89].

---

[2]Recall the distinction between the application developer (the person who customizes the graph editor to the application) and the application programmer (who actually programs the application).

- **Program maintenance:** As shown in figure 1.4, EDGE is used together with a program maintenance program to display import/export relations within Modula programs [Luc90].

- **VLSI design environment:** At Carnegie-Mellon University, EDGE is used in the Demeter project to represent the data flow graph representing the various CAD tools used in developing VLSI applications. EDGE may also be used at the Portland State University as a user interface to the DIADES design automation system for digital circuits. In both cases, EDGE is used to display graphs related to VLSI design (e.g. dataflow graphs) not to perform VLSI layout.

- **Database:** EDGE is used to display generalization hierarchies as part of the DEED project [Seu90, KL88]. Here, the hierarchical subgraph abstractions are used to display information for each object type in the schema. By "zooming in" on an abstraction, the user can view additional information about the object (e.g. its set of attributes).

- **Knowledge representation:** EDGE is used to display a thesaurus as part of a natural language help system based on caseframe parsing [Str90, TAH89]. Several future users are interested in using EDGE to display semantic nets representing knowledge.

Chapter 9

> "*You on the cutting edge of technology have already made yesterday's impossibilities the commonplace realities of today.*"
> – R. Reagan

Summary and Future Research

9.1 Graph Layout

Manual layout of graphs is time-consuming and error-prone. Therefore automatic graph layout is the only reasonable choice for laying out large graphs. However, the user and application developer often want to maintain some control over the layout. Thus, the automatic graph layout algorithm should be able to satisfy layout constraints specified by the user or by the application. The solution presented in section 4.2 achieves this by combining an automatic layout algorithm with a layout constraint manager. Section 4.3 shows how the modified layout algorithm can use layout constraints to produce a stable layout. Not only can the user have a stable layout, but the layout can often be accomplished both faster and with fewer crossings than with an instable layout.

Areas for future research include:

- **Let the system propose when a new layout is worthwhile:** It would be relatively easy to have the system perform the layout calculations in the "background" and suggest to the user when a new layout would be worthwhile. A user-specified parameter could control how much better the layout should be before interrupting the editing session.

- **Allow the user to improve the layout:** So far, the issues of manual and automatic layout have been addressed separately. In fact, a combination of the two is often desirable. One promising approach is to have the system perform an automatic layout of the graph and subsequently allow the user to move individual graph objects. Layout constraints could be used to stabilize the graph layout by recording the user's improved layout of the graph. If the user could express as a constraint why the node was moved to the new position, this could, of course, be reflected in all future layouts as well.

- **More support for the wide variety of "traditional" representations:** One hindrance to automatic graph layout is the divergence of the graph representations in various application areas. For example, in the area of databases, an entity-relationship diagram is often drawn with only horizontal and vertical lines as shown in figure 9.1(a). Petri nets, on the other hand, are typically displayed as shown in figure 9.1(b). Syntax graphs, used to represent the grammar of a language, use yet another different representation (see figure 9.1(c)). It is difficult to have a single layout algorithm which is adaptable enough to satisfy users accustomed to so many different representations. The development of an automatic layout algorithm which puts an even stronger emphasis on layout constraints than the solution proposed here may be required. Perhaps also, as automatic graph layout algorithms become more prevalent, there will be less diversity in the "traditional" representations.

- **Incremental layout:** In some application areas (e.g. hypertext or exploring search trees), it is desirable to display only a subgraph of the entire graph. As the user explores the graph, additional subgraphs should become visible. The following describes how the Sugiyama layout algorithm could be modified to "simulate" such an incremental layout.

  Initially, the preprocessing and barycentric ordering phases would be applied to the entire graph. The finetuning and display, however, would only take the (initially small set of) visible nodes into account. This would have the effect that, as the user explores the graph and more nodes become visible, the finetuning phase would cause the graph to expand to include these additional nodes. The horizontal and vertical relative positions of the nodes would not change because these are determined only in the preprocessing and barycentric ordering phases and thus are never recalculated.

- **Graph partitioning:** With graph partitioning (see [Mes89, MRH89, MG89]), the graph is split into a set of subgraphs based on their connectivity. When the user makes a change in the graph, only the subgraph(s) affected by the change will be laid out. This approach has the additional advantage that the layout of the subgraphs could be calculated in parallel.

- **Improvements to the display of edges:** With the exception of the edge concentration algorithm, the emphasis here thus far has been more

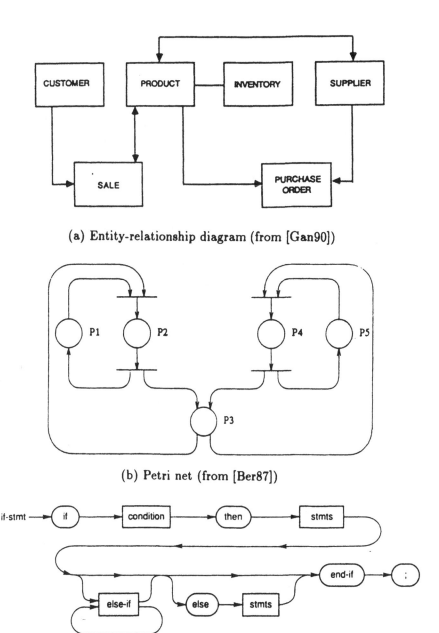

(a) Entity-relationship diagram (from [Gan90])

(b) Petri net (from [Ber87])

(c) Syntax graph (from [Seb89])

Figure 9.1: Traditional representations

on the placement of nodes rather than the edges. The user may well desire
more control over the edges and, in fact, may want to express constraints
on edges as well as on nodes.

The user may prefer that edges be routed around nodes that they intersect,
but these computations can be expensive to do automatically. A graph
editor which supports the combination of automatic and manual layout
could allow the user to manually route individual edges.

A relatively simple addition, which would improve the quality of the layout
significantly, is support for curved edges (splines), loops, and multiple edges
(e.g. see figure 9.1). A placement of the edge labels so that they do not
conflict with the placement of nodes, edges, or other edge labels is also
desirable.

- **Multi-directional layout:** Most of the currently existing layout algorithms do not take different preferred edge orientations into account (with
  the exception of Kb-edit). It is easy to make edges of different types displayed differently (as is done currently in EDGE), but in some cases the
  user wants the layout algorithm to position the nodes based on their edge
  types. Consider, for example, the graph shown in figure 9.2(a). The solid
  edges represent a taxonomy and the dashed edges represent the "is-a" relationship. The user may want one edge type to be displayed horizontally
  and the other vertically to help distinguish between the edge types (as
  shown in figure 9.2(b)). Most layout algorithms, however, treat all edges
  alike. It is possible to restrict the set of edges used in the layout to edges
  of a particular type. But then the other edges must either be drawn in
  ad hoc, which is likely to result in a poor layout, or the user must resort
  to using separate views on the graph, each showing a different edge type.
  Although the additional complexity may slow the layout algorithm, the
  user may still prefer this alternative. Because the graph is displayed in the
  plane, higher orders of layout orientation are not likely to be useful.

9.2   Graphical Abstraction

A user viewing a graph displayed by a graph editor is restricted by two factors
– the physical size of the screen, which may cause parts of the graph which the
user is interested in to be off-screen, and the limited amount of information a
human can digest at once. Abstraction mechanisms help the user group nodes
and edges in a way that makes it easier to understand large graphs.

Several different representations for graphical abstractions were presented:
hierarchical subgraph abstractions, separate views, and edge concentrations.
Particular emphasis was placed on how these graphical abstractions can be defined automatically.

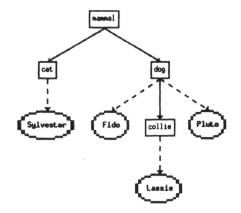

(a) Single-directional layout

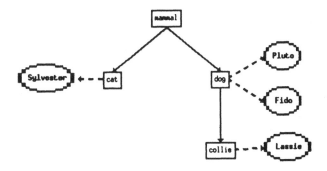

(b) Bi-directional layout

Figure 9.2: Single- and bi-directional layout

Areas for future research include:

- **Other methods for automatically grouping abstractions:** Other methods for automatically defining abstractions should be investigated. In particular, algorithms based on graph-theoretical properties such as strongly connected components are promising. Such techniques are presented in [MG89].

- **Other methods for representing abstractions:** It would be interesting to know more about what abstraction mechanisms users find useful. Hierarchical graphical abstractions are a popular and powerful technique. But, unlike the hierarchical subgraph presented in this dissertation, it may be desirable to allow a node to be in more than one abstraction simultaneously.

- **Let the system propose parameters for the edge concentration:** The system could try to apply a series of edge concentrations on a graph without user intervention. The system could at least suggest which levels and concentration sizes are promising. The choice of which level to concentrate could, for example, be based on the number of crossings between adjacent levels.

- **More reliance on other graphical user interface techniques:** Even within the restricted plane of a bitmapped workstation screen, there are graphical user interface techniques which often go unused. For example, important nodes could be brought to the users attention by animating them (possibly using time as an additional dimension e.g. by flashing them at different speeds). More use could be made of size as an attribute, for example displaying the entire graph by showing nodes which are currently of interest in normal size and the remaining nodes with size inversely proportional to their distance from the "interesting" nodes.

9.3   Persistence

Information relevant to the graph editing session can be saved in a file in the GRL format. This allows the user to restart the editing session at some later time with the same configuration. The format of GRL is designed so that attributes of application-specific graph objects can be saved in the same format as standard attributes.

Areas for future research include:

- **More formatting options within GRL:** One way to further improve GRL would be to allow the user to have more control over the GRL format.

For example, the user can currently specify a label and icon for the node and can control whether the icon is displayed at the top, bottom, or around the node. No matter how much GRL is extended to offer more alternatives, it would be preferable for users to be able to specify their own format. GRL could use a formatting string, similar to the printf format statement in the C language [KR79], to specify the appearance of a node.

- **Printing:** A graph editor is not only useful for interacting with an application, but also for displaying this information as a graph in a document. One of the features EDGE users ask for most is to be able to print a hardcopy of the graph. Although the X Window System has commands to create an X-window dump and to convert it to PostScript, what the user really wants is for the entire graph to be printed, not just what is currently visible in the editing window. After issuing the command, the user should be able to sit back and watch the system automatically scroll the graph by "pages", and print each page. A small amount of overlap should be allowed between the pages to make it easy to join the pieces back together. Since the implementation of GRL used in EDGE already has facilities for scrolling the graph to pre-defined positions, this feature would be easy to add. The system should also be able to produce a PostScript representation of the graph which could be arbitrarily scaled and included in a document.

9.4 Extendibility

Object-oriented design of a graph editor is a natural way to achieve extendibility. The proposed solution shows how object-oriented programming and program generation can be used in concert to achieve extendibility of a system automatically. One of the novel aspects of the program generator tool is that its only input specification is the class declarations themselves – no separate specification by the application developer is necessary.

Areas for future research include:

- **Structure as a toolkit:** One of EDGE's strengths is the set of choices that the user can choose from – for example that the user can choose between several different graph layout algorithms or graphical abstractions. Currently, the code for each of these choices is linked together with EDGE. It would be desirable for the application developer to be able to selectively choose what options should be included in the editor by loading the appropriate libraries. The additional structure would also make it easier for application developers to provide their own layout, abstraction, or persistence mechanisms.

- **More inheritance among graph objects:** In retrospect it would have been possible to take more advantage of similarities between nodes and edges. One could have defined a class for "graph object" whose methods are used by both nodes and edges.

- **Avoid recompilation:** The possibility of dynamically loading the application's object code into the address space of the graph editor should be investigated. This would allow the application developer to avoid the recompilation step that often occurs when adapting EDGE to a new application.

- **Library of graph routines:** A set of graph algorithms (e.g. to determine shortest path, strongly connected components, articulation points etc.) would be a welcome addition. These should, however, be available as a library, rather than being built-in to the graph editor, because this allows the application developer to load only those routines that are needed.

9.5   Other Areas

Future research in graph editors should also investigate the following topics:

- **Further customization of the user interface:** The user can control the list of entries in the graph, node, and edge menus by using the program generator tool AGENT, but the rest of the user interface of the graph editor is fixed. It would be nice if the user could customize the user interface to a greater extent. Unless this is provided, there is a danger that application developers will disassemble EDGE and use only portions of it in their application. However, care must be taken to balance the benefits of user interface customization against the benefits of having a consistent user interface across applications.

- **Undo and history capabilities:** Any editing system benefits from an undo capability and a graph editor is no exception. Some of the individual solutions proposed here have undo facilities (e.g. the user can undo a layout constraint and the user can dissolve a subgraph abstraction), but what is lacking is a general-purpose undo facility. For example, if the user dissolves a subgraph abstraction or closes a separate view, there are no facilities to easily recreate them. It may also be useful to the user, especially after a long series of edits, to review (and possibly reissue) previous editing commands.

  One potential solution is to base an undo and history mechanism on the graph representation language GRL. Although, in theory, one could save a GRL representation of the editing session after every editing change, a more practical approach would be to save a more compact representation (possibly saving a "delta" of the graph representations between checkpoints).

- **Independent execution:** The solutions presented thus far assume that an application developer is willing (and able) to compile the application code together with the graph editor. Consideration should be given to the possibility of running the graph editor and the application as separate processes. The application could communicate with the graph editing process (possibly even running on a remote machine). Due to its modularity, the graph representation language GRL is a natural candidate as the basis for this communication. For the initial display of a graph either a GRL specification of the graph could be sent or a message telling the server what file to find it in. During the editing session itself, portions of the GRL file (e.g. the specification for a single node or edge) could be sent.

Chapter 10

Conclusion

Graph editors are a powerful and widely-applicable tool because they combine a general graphical representation of information (a graph) with a general model of interaction (an editor). An extendible graph editor provides an effective and natural user interface to applications because it allows the user to manipulate objects associated with the application directly. The wide range of applications shown in the examples presented here represent a fraction of the possible application areas for an extendible graph editor. The problem areas addressed – graph layout, graphical abstraction, persistence, and extendibility – are common problems for all graph editors. Solutions for each of these fundamental and recurring problems associated with graph editors were presented here.

The goal was to design a graph editor which is easy to adapt to many different applications. The advantages of having a graphical interface to the application will thus be available to the application developer for a minimal customization effort. One of the ways to reduce the effort of the application developer is object-oriented design, because this allows the application developer to concentrate on application-specific aspects. A second way to reduce the effort is reliance on automation. Automation is a common theme to all of the proposed solutions.

- The automatic graph layout algorithm is enhanced so that it can automatically take layout constraints specified by the user or by the application into account.

- A set of low-priority layout constraints, recording the current layout, are automatically generated before a new layout is performed. The layout algorithm can use these constraints to keep the layout stable.

- The subgraph abstractions can be defined automatically. Several grouping mechanisms (e.g. all objects of a particular type, all objects whose labels match a regular expression etc.) are built-in to the graph editor. Alternatively, the application developer or user can provide a function which defines the subgraph.

- The edge concentration algorithm provides a clustering mechanism for automatically grouping edges together in order to make the graph more readable.

- Persistence is achieved through a program generator tool which automatically generates large portions of the code necessary for integrating EDGE with a new application. In particular, the input, output, and menu code is extended to take attributes specified in the application's class declarations into account. Basic constructors and destructors for the application's classes are also generated.

One of the advantages of automation is that it saves effort for both the user and the application developer. The user benefits from having an automatic layout, and in particular from having a stable layout. The application developer benefits from the generation of source code which takes application-specific attributes automatically into account. In both cases, automation relieves a human of tedious and time-consuming work.

A second advantage of automation is that it can be less prone to errors because the results are produced automatically from a relatively simple specification. For example, in the case of the program generator tool AGENT, the application developer does not even have to provide a separate specification – the code is generated directly from an existing piece of software (the class declarations). A related advantage is that the result is more consistent and therefore easier to maintain than the same code produced manually.

The solutions proposed in this thesis have shown that the benefits of automation can be obtained for a low cost. Recall that the stable layouts were not only stable, but were often determined faster (and with fewer crossings) than their instable counterparts. The edge concentration algorithm made dramatic reductions in the numbers of crossings in a graph, thus making the graph more readable, in a matter of seconds. The program generator tool produces code directly from the class declarations, so that the generated code can be obtained with almost no effort on the part of the application developer.

Automatic techniques are not, however, yet powerful enough to be unleashed without user control. The user often wants to be able to influence the extent of the automation. In many cases, the automatic solution simply cannot be expected to "know" the necessary information because this information may depend on the semantics of the application. For example, recall that the justification for offering layout constraints was that most automatic layout algorithms concern themselves solely with the structure and not the semantics of the graph. Similarly, the program generator tool cannot be expected to know that certain fields of the data structure are for internal use only and should not be offered in the menu – therefore the application developer must specify this information through the exception list.

An automated solution which works under control of the user is the most promising approach for a radical reduction of effort when developing an extendible graphical user interface for an application. The availability of an extendible graph editor which incorporates the solutions proposed here will make the use of graph-based graphical user interfaces more widespread.

Appendix A

EBNF Grammar for GRL

Lexical Conventions

The lexical conventions used are as follows. Terminals are shown in bold font. Non-terminals are shown in angle brackets (<>). Optional parts are shown in square brackets ([]). Parts that can be repeated or omitted are shown in parenthesis. A "*" following the set of parenthesis indicates zero or more repetitions. Comments in C or in C++ style can be present anywhere in the GRL specification.

Grammar

The following gives the grammar for the standard version of GRL. The entries <*graph attribute*>, <*node attribute*>, and <*edge attribute*> can be extended by using the program generator tool AGENT.

<*graph*>	:=	**graph :**{ (<*graph entry*>)* }**.**
<*graph entry*>	:=	<*graph attribute*>
	|	<*node defaults*>
	|	<*edge defaults*>
	|	<*node*>
	|	<*edge*>
	|	<*constraint*>**.**
<*graph attribute*>	:=	<*graph attribute name*> **:** <*attribute value*>**.**

\<graph attribute name\>	:=	any attribute name shown in table 6.1.
\<node defaults\>	:=	**node.***\<node attribute\>.*
\<edge defaults\>	:=	**edge.***\<edge attribute\>.*
\<node\>	:=	**node:** $\{$ (*\<node attribute\>*)* $\}$.
\<edge\>	:=	**edge:** $\{$ (*\<edge attribute\>*)* $\}$.
\<constraint\>	:=	**constraint:** $\{$ (*\<constraint attribute\>*)* $\}$.
\<node attribute\>	:=	*\<node attribute name\>* **:** *\<attribute value\>.*
\<edge attribute\>	:=	*\<edge attribute name\>* **:** *\<attribute value\>.*
\<constraint attribute\>	:=	*\<constraint attribute name\>* **:** *\<attribute value\>.*
\<node attribute name\>	:=	any attribute name shown in table 6.2.
\<edge attribute name\>	:=	any attribute name shown in table 6.3.
\<constraint attribute name\>	:=	any attribute name shown in table 6.4.
\<attribute value\>	:=	*\<integer value\>*
	|	*\<float value\>*
	|	*\<character value\>*
	|	*\<string value\>*
	|	*\<struct value\>*
	|	*\<array value\>*
	|	*\<enum value\>.*
\<integer value\>	:=	any integer constant in C style.
\<float value\>	:=	any floating constant in C style.
\<character value\>	:=	'*\<character\>* '.
\<string value\>	:=	"(*\<character\>*)* ".
\<struct value\>	:=	$\{$ (*\<attribute\>*)* $\}$.
\<array value\>	:=	$\{$ (*\<index value\>*)* $\}$.
\<enum value\>	:=	*\<enum constant\>.*

\<enum constant\>	:=	*\<letter\>* (*\<letter\>* | *\<digit\>*)*.
\<index value\>	:=	*\<attribute value\>*
	|	*\<index\>* : *\<atribute value\>*
	|	*\<range\>* : *\<atribute value\>*
	|	***:** *\<atribute value\>*.
\<range\>	:=	[*\<integer value\>* , *\<integer value\>*].
\<index\>	:=	*\<integer value\>*.
\<character\>	:=	any ASCII character.
\<letter\>	:=	**a** | **b** | ... | **z** | **A** | ... | **Z** | **_** .
\<digit\>	:=	**0** | **1** | ... | **9**.

Bibliography

[ABC+83]   M.P. Atkinson, P.J. Bailey, K.J. Chisholm, W.P. Cockshott, and R. Morrison. An approach to persistent programming. *The Computer Journal*, 26(4):360–365, 1983.

[ACCM83]   Malcolm Atkinson, Ken Chisholm, Paul Cockshott, and Richard Marshall. Algorithms for a persistent heap. *Software—Practice and Experience*, 13(3):259–271, March 1983.

[Ase87]   Paul J. Asente. Editing graphical objects using procedural representations. Technical report, Digital Equipment Corporation, November 1987. Revised version of author's PhD dissertation from Stanford University.

[BCL90]   Duane A. Bailey, Janice E. Cuny, and Craig P. Loomis. ParaGraph: Graph editor support for parallel programming environments. *International Journal of Parallel Programming*, 19(2), April 1990.

[BL76]   K. Booth and G. Lueker. Testing for the consecutive ones property, interval graphs, and graph planarity using PQ-tree algorithms. *Journal of Computer and System Sciences*, 13:335–379, 1976.

[BM92]   A.L. Brown and R. Morrison. A generic persistent object store. *Software Engineering Journal*, pages 161–168, March 1992.

[BMS86]   D.G. Bobrow, S. Mittal, and M.J. Stefik. Expert systems: Perils and promises. *Communications of the ACM*, 29(9):880–894, 1986.

[BNT86]   Carlo Batini, Enrico Nardelli, and Roberto Tamassia. A layout algorithm for data flow diagrams. *IEEE Transactions on Software Engineering*, SE-12(4):538–546, April 1986.

[BNTT85]   C. Batini, E. Nardelli, M. Talamo, and R. Tamassia. GINCOD: A graphical tool for conceptual design of data base applications. In V. De Antonellis A. Albano and A. Di Leva, editors, *Computer-Aided Database Design: The DATAID Project*, chapter II. Elsevier Science Publishers B.V. (North-Holland), 1985.

[Böh89]   Karl-Friedrich Böhringer.   Stabilität von Algorithmen für
            Graphenumbruch (Stability in graph layout algorithms). Master's
            thesis, University of Karlsruhe, Department of Informatics, July
            1989.

[Bor79]   Alan H. Borning. *ThingLab - A Constraint-Oriented Simulation
            Laboratory.* PhD thesis, Stanford University, Department of Com-
            puter Science, March 1979. (Xerox PARC Rep. SSL-79-3).

[BP90]   Karl-Friedrich Böhringer and Frances Newbery Paulisch. Using con-
            straints to achieve stability in automatic graph layout algorithms.
            In *Proceedings of the SIGCHI (Special Interest Group Computer
            Human Interaction) '90*, Seattle, WA, April 1-5 1990.

[Bra79]   Ronald J. Brachman. On the epistemological status of semantic
            networks. In Nicholas V. Findler, editor, *Associative Networks -
            Representation and Use of Knowledge by Computers*, chapter 1.
            Academic Press, 1979.

[Bro88]   Kenneth P. Brooks. *A Two-view Document Editor with User-
            definable Document Structure.* PhD thesis, Stanford University, De-
            partment of Computer Science, November 1988.

[Bru88]   Glenn Bruns. Germ: A metasystem for browsing and editing. Tech-
            nical Report MCC Technical Report STP-122-88, MCC/Software
            Technology Program, Austin, TX, April 12 1988.

[BS89]   Bjarni Birgisson and Gregory E. Shannon. GraphView: An exten-
            sible interactive platform for manipulating and displaying graphs.
            Technical Report 295, Computer Science Department, Indiana Uni-
            versity, Indiana University, Department of Computer Sciences,
            Bloomington, IN, December 1989.

[BT88]   G. Di Battista and R. Tamassia. Algorithms for plane representa-
            tion of acyclic digraphs. *Theoretical Computer Science*, 61:175–198,
            1988.

[Car80]   Marie-Jose Carpano. Automatic display of hierarchized graphs for
            computer-aided decision analysis. *IEEE Transactions on Systems,
            Man, and Cybernetics*, SMC-10(11):705–715, November 1980.

[CCP87]   C.B. Chapman, D.F. Cooper, and M.J. Page. *Management for
            Engineers.* John Wiley and Sons, 1987.

[Che76]   Peter Pin-Shan Chen. The entity-relationship model — toward
            a unified view of data. *ACM Transactions on Database Systems*,
            1(1):9–36, March 1976.

[CL88]   Ching-Hua Chow and Simon S. Lam. PROSPEC: An interac-
            tive programming environment for designing and verifying commu-
            nication protocols. *IEEE Transactions on Software Engineering*,
            14(3):327–338, March 1988.

[Cle88]   J. Craig Cleaveland. Building Application Generators. *IEEE Soft-
            ware*, 5(4):25–33, July 1988.

[CM88]   Luca Cardelli and David MacQueen. Persistence and type abstrac-
            tion. In Malcolm P. Atkinson, Peter Buneman, and Ronald Morri-
            son, editors, *Data Types and Persistence*, chapter 3. Springer Verlag,
            1988.

[CON85]   N. Chiba, K. Onoguchi, and T. Nishizeki. Drawing plane graphs
            nicely. *Acta Informatica*, 22:187–201, 1985.

[Con87]   Jeff Conklin. Hypertext: An introduction and survey. *IEEE Com-
            puter*, 20(9):17–41, September 1987.

[Cox86]   B. J. Cox. *Object-Oriented Programming: An Evolutionary Ap-
            proach*. Addison-Wesley, Reading, MA, 1986.

[Dav86]   Michael Davis. A layout algorithm for a graph browser. Technical
            report, Computer Science Division, University of California, Berke-
            ley, 1986.

[Dav87]   Ernest Davis. Constraint propagation with interval labels. *Artificial
            Intelligence*, 32:281–331, 1987.

[DC]   Ian Darwin and Geoff Collyer. Softquad Inc., Private communica-
            tion.

[DK91]   David Dobkin and Eleftherios Koutsofios. LEFTY: A two-view ed-
            itor for technical pictures. In *Graphics Interface '91*, pages 68–76,
            1991.

[DS90]   Prasun Dewan and Marvin Solomon. An approach to support au-
            tomatic generation of user interfaces. *ACM Transactions on Pro-
            gramming Languages and Systems*, 12(4):566–609, October 1990.

[Ead84]   P. Eades. A heuristic for graph drawing. *Congressus Numeratium*,
            42:149–160, 1984.

[ENRE87]   H. Ehrig, M. Nagl, G. Rozenberg, and A Rosenfeld (Eds). Graph-
            grammars and their application to computer science, lecture notes
            in computer science no. 291. Berlin, 1987. Springer Verlag.

[ES90]   Peter Eades and Kozo Sugiyama. How to draw a directed graph.
            Journal of Information Processing, 13(4):424–437, 1990.

[ET89]   Peter Eades and Roberto Tamassia. Algorithms for automatic graph
           drawing: An annotated bibliography. Technical Report CS-89-09
           (Revised Version), Brown University, Department of Computer Sci-
           ence, Providence, RI, October 1989.

[Eve79]   Shimon Even. *Graph Algorithms*. Computer Science Press,
           Rockville, MD, 1979.

[EW89]   P. Eades and N. C. Wormald. Edge crossings in drawings of bipartite
           graphs. Technical Report 109, University of Queensland, Dept. of
           Computer Science, 1989.

[Far48]   I. Fary. On straight line representation of planar graphs. *Acti Sci.
           Math. Szegd*, 11:229–233, 1948.

[FBMB90]   Bjorn N. Freeman-Benson, John Maloney, and Alan Borning. An in-
           cremental constraint solver. *Communications of the ACM*, 33(1):54–
           63, January 1990.

[Fel79]   Stuart I. Feldman. Make — a program for maintaining computer
           programs. *Software—Practice and Experience*, 9(3):255–265, March
           1979.

[For71]   J. W. Forrester. *World Dynamics*. Wright-Allen Press, Cambridge,
           MA, 1971.

[FR91]   Thomas M. J. Fruchterman and Edward M. Reingold. Graph dr-
           rawing by force-directed placement. *Software—Practice and Expe-
           rience*, 21(11):1129–1164, November 1991.

[Gan90]   Chris Gane. *Computer-Aided Software Engineering - the method-
           ologies, the products, and the future.* Prentice Hall, Englewood
           Cliffs, NJ, 1990.

[GJ79]   Michael R. Garey and David S. Johnson. *Computers and Intractabil-
           ity — A Guide to the Theory of NP-Completeness.* W. H. Freeman
           and Company, San Francisco, CA, 1979.

[GJ83]   M. R. Garey and D. S. Johnson. Crossing number is NP-Complete.
           SIAM Journal of Algebraic and Discrete Methods, 4(3):312–316,
           September 1983.

[GKNV]   Emden R. Gansner, Eleftherios Koutsofios, Stephen C. North, and
           Kiem-Phong Vo. A technique for drawing directed graphs. *IEEE
           Transactions on Software Engineering.* to appear.

[Gli90]   Ephraim P. Glinert, editor. *Visual Programming Environments -
           Applications and Issues.* IEEE Computer Society Press, Los Alami-
           tos, CA, 1990.

[GNV88]   E.R. Gansner, S. C. North, and K. P. Vo. DAG: A program
          that draws directed graphs. *Software—Practice and Experience*,
          18(11):1047–1062, November 1988.

[Gol84]   Adele Goldberg. *Smalltalk-80 - The Interactive Programming En-
          vironment*. Addison-Wesley, Reading, MA, 1984.

[Him88]   Michael Himsolt. Entwicklung eines Grapheneditors (Design of a
          graph editor). Master's thesis, Universität Passau, Fakultät für
          Mathematik und Informatik, June 1988.

[HSF90]   Jens Herrmann, Andreas Spillner, and Reinhold Frank. Pitfall –
          Ein Werkzeug für den Integrationstest (Pitfall – an integration test-
          ing tool). Technical Report PIT-Report 90/2, Universität Bremen,
          Fachbereich Mathematik/Informatik, February 1990.

[HT74]   John Hopcroft and Robert Tarjan. Efficient planarity testing. *Jour-
          nal of the ACM*, 21(4):549–568, 1974.

[JG89]   David Jablonsowski and Vincent A. Guarna, Jr. GMB: A tool
          for manipulating and animating graph data structures. *Software—
          Practice and Experience*, 19(3):283–301, March 1989.

[Joh75]   S. C. Johnson. YACC: Yet Another Compiler Compiler. *Bell Lab-
          oratories Computing Science Technical Report 32*, July 1975.

[Joh82]   David S. Johnson. The NP-completeness column: An ongoing guide.
          Journal of Algorithms, 0(3):89–99, March 1982.

[JRV+89] Jeff Johnson, Teresa L. Roberts, William Verplank, David C. Smith,
          Charles H. Irby, Marian Beard, and Kevin Mackey. The Xerox Star:
          A retrospective. *IEEE Computer*, 22(9):11–29, September 1989.

[Ker82]   Brian W. Kernighan. *PIC - A Graphics Language for Typesetting
          User Manual*, 1982.

[KK89]   T. Kamada and S. Kawai. An algorithm for drawing general undi-
          rected graphs. *Information Processing Letters*, 31:7–15, 1989.

[KL88]   S. Karl and P. C. Lockemann. Design of engineering databases:
          A case for more varied semantic modelling concepts. *Information
          Systems*, 13:335–357, 1988.

[KN91]   Eleftherios Koutsofios and Stephen C. North. *DOT User's Manual:
          Drawing Graphs with DOT*. AT&T Bell Laboratories, Murray Hill,
          NJ, September 1991.

[Knu71]   D.E. Knuth. Optimum binary search trees. *Acta Informatica*, 1:14–
          25, 1971.

[KR79]   Brian W. Kernighan and Dennis M. Ritchie. *The C Programming Language*. Prentice Hall, Englewood Cliffs, NJ, 1979.

[Lam87]   David Alex Lamb. IDL: Sharing intermediate representations. *ACM Transactions on Programming Languages and Systems*, 9(3):297–318, July 1987.

[LB86]   P. A. Luker and A. Burns. Program Generators and Generation Software. *The Computer Journal*, 29(4):315–321, 1986.

[Lel88]   William Leler. *Constraint Programming Languages – Their Specification and Generation*. Addison-Wesley, Reading, MA, 1988.

[LNS85]   R. J. Lipton, S. C. North, and J. S. Sandberg. A method for drawing graphs. In *Proceedings of the Symposium on Computational Geometry*, pages 153–160, Baltimore, MD, June 1985. ACM.

[LS75]   M. E. Lesk and E. Schmidt. LEX - A Lexical Analyser Generator. *Bell Laboratories Computing Science Technical Report 39*, October 1975.

[Luc90]   Jürgen Lucas. Ein Informationssystem für die Wartung großer Programme (An information system for maintenance of large progams). University of Karlsruhe, Department of Informatics, December 1990.

[MA86]   Joseph Manning and Mikhail J. Atallah. Fast detection and display of symmetry in trees. Technical Report TR-606, Purdue University, Department of Computer Sciences, W. Lafayette, IN, 1986.

[Man90a]   Stefan Manke. Generierung von Graphbearbeitungsprogrammen aus objekt-orientierten Spezifikationen (Generation of graph manipulation programs from object-oriented specifications). Master's thesis, University of Karlsruhe, Department of Informatics, March 1990.

[Man90b]   Joseph Brendan Manning. *Geometric Symmetry in Graphs*. PhD thesis, Purdue University, Department of Computer Sciences, W. Lafayette, IN, December 1990.

[MBFB89]   John H. Maloney, Alan Borning, and Bjorn N. Freeman-Benson. Constraint technology for user-interface construction in ThingLab II. In *Proc. of the OOPSLA (Object-Oriented Programming Systems, Languages, and Applications) '89*, pages 381–388, October 1989.

[Mes89]   Eli Benjamin Messinger. *Automatic Layout of Large Directed Graphs*. PhD thesis, University of Washington, Department of Computer Sciences, July 1989. TR Number 88-07-08.

[Mey87]   Bertrand Meyer. EIFFEL: Programming for reusability and ex-
           tendibility. *SIGPLAN Notices*, 22(2):85–94, February 1987.

[Mey88]   Bertrand Meyer. *Object-oriented Software Construction*. Prentice
           Hall, Englewood Cliffs, NJ, 1988.

[MG89]   Thomas P. Murtagh and Douglas J. Gschwind. A recursive algo-
           rithm for drawing hierarchical directed graphs. Technical Report
           No. CS-89-02, Williams College, Williamstown, MA, June 7 1989.

[MRH89]   Eli B. Messinger, Lawrence A. Rowe, and Robert H. Henry. A
           divide-and-conquer algorithm for the automatic layout of large di-
           rected graphs. Technical report, University of California at Berkeley,
           Computer Science Division – EECS Department, February 1989.
           UCB/ERL M89/23.

[Mye87]   Brad A. Myers. *Creating user interfaces by demonstration*. PhD
           thesis, University of Toronto, Department of Computer Sciences,
           1987. (Academic Press, 1988).

[Mye90]   Brad A. Myers. Creating user interfaces using programming by ex-
           ample, visual programming, and constraints. *ACM Transactions on
           Programming Languages and Systems*, 12(2):143–177, April 1990.

[Nel85]   G. Nelson. Juno, a constraint-based graphics system. In B. Barsky,
           editor, *SIGGRAPH Computer Graphics*, pages 235–243, San Fran-
           cisco, July 1985.

[New88]   Frances J. Newbery. An interface description language for graph
           editors. In *Proc. of the IEEE 1988 Workshop on Visual Languages*,
           Pittsburgh, PA, October 1988.

[New89]   Frances J. Newbery. Edge concentration: A method for clustering
           directed graphs. In *Proc. of the Second International Workship on
           Software Configuration Management*, Princeton, NJ, October 24-27
           1989.

[NNGS90]   John R. Nestor, Joseph M. Newcomer, Paola Giannini, and Don-
           ald L. Stone. *IDL: The Language and its Implementation*. Prentice
           Hall, Englewood Cliffs, NJ, 1990.

[OvW78]   R.H.J.M. Otten and J.G. van Wijk. Graph representations in inter-
           active layout design. In *Proceedings IEEE International Symposium
           on Circuits and Systems*, pages 914–918, New York, NY, 1978.

[Par72]   David L. Parnas. On the criteria to be used in decomposing sys-
           tems into modules. *Communications of the ACM*, 15(2):1053–1058.

[PK86]   Jeff Pepper and Gary Kahn. Knowledge Craft: An environment
           for rapid prototyping of expert systems. In *Proceedings of the SME
           Conference on Artificial Intelligence for the Automotive Industry.*
           SME, March 1986.

[PM90]   Frances Newbery Paulisch and Stefan Manke.   Automated ex-
           tendibility in an object-oriented system. November 1990.

[PMT90]   Frances Newbery Paulisch, Stefan Manke, and Walter F. Tichy.
           Persistence for arbitrary C++ data structures. In *Proc. of Inter-
           national Workshop on Computer Architectures to Support Security
           and Persistence of Information*, Bremen, FRG, May 8-11 1990.

[RC89]   Joel E. Richardson and Michael J. Carey. Persistence in the E
           language: Issues and implementation. *Software—Practice and Ex-
           perience*, 19(12):1115–1150, December 1989.

[RDLK90]   Vaclav Rajlich, Nicholas Damaskinos, Panagiotis Linos, and Wafa
           Khorshid. VIFOR: A tool for software maintenance. *Software—
           Practice and Experience*, 20(1):66–77, January 1990.

[RDM+87]   Lawrence A. Rowe, Michael Davis, Eli Messinger, Carl Meyer,
           Charles Spirakis, and Allen Tuan. A browser for directed graphs.
           Software—Practice and Experience, 17(1):61–76, January 1987.

[RM88]   M. Reggiani and F. Marchetti. A proposed method for representing
           hierarchies. *IEEE Transactions on Systems, Man, and Cybernetics*,
           18(1):2–8, January 1988.

[Rob87]   Gabriel Robins. The ISI grapher: a portable tool for displaying
           graphs pictorially. Helsinki, Finland, August 17-18 1987. Symbol-
           iikka '87. Also Technical Report IST/RS-87-196 Information Sci-
           ences Institute, Marina Del Rey, CA.

[RT81]   Edward M. Reingold and John S. Tilford. Tidier drawings of trees.
           IEEE Transactions on Software Engineering, 7(2):223–228, March
           1981.

[Rum89]   Bernd Rumscheid. Ein Umbruchalgorithmus mit Schwerpunkt auf
           einer planaren Darstellung (A planar layout algorithm). Undergrad-
           uate Thesis, University of Karlsruhe, Department of Informatics,
           July 1989.

[Sca89]   Richard Snodgrass and contributing authors. *The Interface De-
           scription Language: Definition and Use.* Computer Science Press,
           Rockville, MD, 1989.

[Seb89]   Robert W. Sebesta. *Concepts of Programming Languages.* Ben-
           jamin/Cummings Publishing Company, Redwood City, CA, 1989.

[Seu90]   Hans-Ulrich Seufert. Interaktive Transformation komplexer Gener-
          alisierungshierachien durch Kombination partiell anwendbarer Ver-
          fahren (Interactive transformation of complex generalization hier-
          archies through combination of partially applicable methods). Mas-
          ter's thesis, University of Karlsruhe, Department of Informatics,
          September 1990.

[SG86]   Robert W. Scheifler and Jim Gettys. The X window system. *ACM
          Transactions on Graphics*, 5(2):79–109, April 1986.

[Shn83]   Ben Shneiderman. Direct manipulation: A step beyond program-
          ming languages. *IEEE Computer*, 16(8):57–69, August 1983.

[Shu89]   N.C. Shu. *Visual Programming*. Van Nostrand Reinhold, New York,
          1989.

[SM88]   Pedro A. Szekely and Brad A. Myers. A user interface toolkit based
          on graphical objects and constraints. In *Proc. of the OOPSLA
          (Object-Oriented Programming Systems, Languages, and Applica-
          tions) '88*, pages 36–45, September 1988.

[SM91]   Kozo Sugiyama and Kazuo Misue. Visualization of structural infor-
          mation: Automatic drawing of compund digraphs. *IEEE Transac-
          tions on Systems, Man, and Cybernetics*, 21(4):1–17, July/August
          1991.

[Sny80]   W. Van Snyder. Compact storage of sparse graphs and sets of sets
          with application to LR parser decision functions. Computing Mem-
          orandum Computing Memorandum 468, Jet Propulsion Laboratory,
          Pasadena, CA, September 1980.

[SP89]   Robert W. Schwanke and Michael A. Platoff. Cross references are
          features. In *Proceedings of the Second International Workshop on
          Software Configuration Management*, pages 86–95, Princeton, NJ,
          October 24 1989.

[SS78]   J. Sutton and R. Sprague. A study of display generation and man-
          agement in interactive business applications. Technical Report Tech.
          Rep. RJ2392(31804), IBM San Jose Research Laboratory, Novem-
          ber 1978.

[Str86]   Bjarne Stroustrup. *The C++ Programming Language*. Addison-
          Wesley, Reading, MA, 1986.

[Str90]   Stefan Strugies. Entwurf und Implementierung einer graphischen
          Schnittstelle zur Entwicklung natürlichsprachlicher Grammatiken
          (Design and implementation of a graphical interface for natural lan-
          guage grammars). Undergraduate Thesis, University of Karlsruhe,

[STT81]   Kozo Sugiyama, Shojiro Tagawa, and Mitsuhiko Toda. Methods for visual understanding of hierarchical system structures. *IEEE Transactions on Systems, Man, and Cybernetics*, SMC-11(2):109–125, February 1981.

[Sug84]   Kozo Sugiyama. A readability requirement in drawing digraphs: Level assignment and edge removal for reducing total length of lines. Technical Report 45, International Institute for Advanced Study of Social Information Science, Numazu, Japan, March 1984.

[Sut63]   I. E. Sutherland. Skethpad: A man-machine graphical communication system. In *Proc. of the AFIPS Spring Joint Computer Conference*, pages 329–345, 1963.

[Sys85]   Adobe Systems. *PostScript Language Reference Manual*. Reading, MA, 1985.

[TAH89]   Walter F. Tichy, Rolf L. Adams, and Lars Holter. NLH/E: A natural language help system. In *Proc. of the 11th International Conference on Software Engineering*, Pittsburgh, PA, May 1989.

[Tam87]   Roberto Tamassia. On embedding a graph in the grid with the minimum number of bends. *SIAM Journal on Computing*, 16(3), 1987.

[TBB89]   Roberto Tamassia, Guiseppe Di Battista, and Carlo Batini. Automatic graph drawing and readability of diagrams. *IEEE Transactions on Systems, Man, and Cybernetics*, SMC-18(1):61–79, Jan/Feb 1989.

[TBT83]   R. Tamassia, C. Batini, and M. Talamo. An algorithm for automatic layout of entity relationship diagrams. In C.G. Davis, S. Jajodia, P.A. Ng, and R.T. Yeh, editors, *Entity-Relationship Approach to Software Engineering*, pages 421–439. North-Holland Publishing Co, 1983.

[Tri88]   Howard Trickey. DRAG: A graph drawing system. In *International Conference on Electronic Publishing, Document Manipulation, and Typography*, pages 171–182, Nice, France, April 1988. Cambridge University Press.

[TT89]   R. Tamassia and I.G. Tollis. Planar grid embedding in linear time. *Proceedings IEEE International Symposium on Circuits and Systems*, CAS-36(9):1230–1234, 1989.

[Tut63]   W. T. Tutte. How to draw a graph. *Proceedings of the London Mathematical Soceity*, 3(13):743–768, 1963.

[TW87]   Walter F. Tichy and Blake Ward. A knowledge-based graphical
            editor. Technical Report 3/87, University of Karlsruhe, Department
            of Informatics, January 1987.

[vHE87]   Frans van Hoeve and Rolf Engmann. An Object-oriented Approach
            to Application Generation. *Software—Practice and Experience*,
            17(9):623–645, September 1987.

[vW82]   C.J. van Wyk. A high-level language for specifying pictures. *ACM
            Transactions on Graphics*, 1(2):163–182, April 1982.

[War77]   John N. Warfield. Crossing theory and hierarchy mapping. *IEEE
            Transactions on Systems, Man, and Cybernetics*, SMC-7(7):505–
            523, July 1977.

[Wil83]   Gregg Williams. The LISA computer system. *Byte*, pages 33–50,
            February 1983.

[Wir76]   N. Wirth. *Algorithms + Data Structures = Programs*. Prentice
            Hall, Englewood Cliffs, NJ, 1976.

[Woo81]   Donald R. Woods. *Drawing Planar Graphs*. PhD thesis, Stanford
            University, Department of Computer Science, June 1981. Report
            No. STAN-CS-82-943.

[WP86]   Anthony I. Wasserman and Peter A. Pircher. A graphical, extensible
            integrated environment for software development. In Peter Hender-
            son, editor, *Proceedings of the ACM SIGSOFT/SIGPLAN Software
            Engineering Symposium on Practical Software Development Envri-
            onments*, pages 131–142, Palo Alto, CA, March 1986.

[WS79]   C. Wetherell and A. Shannon. Tidy drawings of trees. *IEEE Trans-
            actions on Software Engineering*, 5(5):514–520, September 1979.

[WWFT88]   Jack C. Wileden, Alexander L. Wolf, Charles D. Fisher, and Peri L.
            Tarr. PGRAPHITE: An experiment in persistent typed object man-
            agement. In Peter Henderson, editor, *Proceedings of the ACM SIG-
            SOFT/SIGPLAN Software Engineering Symposium on Practical
            Software Development Environments (Software Engineering Notes
            Vol. 13, No. 5 or Sigplan Notices Vol. 24, No. 2)*, pages 130–142,
            Boston, MA, Nov. 28-30 1988.

Index

Springer-Verlag
and the Environment

We at Springer-Verlag firmly believe that an international science publisher has a special obligation to the environment, and our corporate policies consistently reflect this conviction.

We also expect our business partners – paper mills, printers, packaging manufacturers, etc. – to commit themselves to using environmentally friendly materials and production processes.

The paper in this book is made from low- or no-chlorine pulp and is acid free, in conformance with international standards for paper permanency.

Printing: Weihert-Druck GmbH, Darmstadt
Binding: Buchbinderei Schäffer, Grünstadt

Lecture Notes in Computer Science

For information about Vols. 1–629
please contact your bookseller or Springer-Verlag